Strategies for Success in the New Telecommunications Marketplace

For a complete listing of the *Artech House Telecommunications Library*,
turn to the back of this book.

Strategies for Success in the New Telecommunications Marketplace

Karen G. Strouse

Artech House
Boston • London
www.artechhouse.com

Library of Congress Cataloging-in-Publication Data
Strouse, Karen G.
Strategies for success in the new telecommunications marketplace / Karen G. Strouse.
p. cm.—(Artech House telecommunications library)
Includes bibliographical references and index.
ISBN 1-58053-142-3 (alk. paper)
1. Telecommunication—Management. 2. Competition, International.
I. Title. II. Series.
HE7631 .S76 2000
384'.068'8—dc21 00-048093
CIP

British Library Cataloguing in Publication Data
Strouse, Karen G.
Strategies for success in the new telecommunications marketplace.
— (Artech House telecommunications library)
1. Telecommunication—Management 2. Competition, International
I. Title
384'.068
ISBN 1-58053-142-3

Cover design by Gary Ragaglia

685 Canton Street
Norwood, MA 02062

International Standard Book Number: 1-58053-142-3
Library of Congress Catalog Card Number: 00-048093

10 9 8 7 6 5 4 3 2 1

To Arthur

Contents

Preface

Deregulation of the telecommunications industry is in progress almost everywhere in the world. It is still too soon to declare any winners—or losers—but the game is at least under way, and the game plan is becoming visible.

Strategic challenges are everywhere. How big does a telecommunications service provider have to be to succeed, and when will competitors stop growing even larger? How should a service provider decide where to land on the distribution chain and how widely to spread out? Is it necessary to control the entire business in-house, owning all facilities, managing information systems, or supporting customers? How can telecommunications service providers break free of the commodity status that causes price wars and a level of churn that surpasses nearly every other industry? Why did some strategies succeed and others fail?

This book examines the strategic approaches of telecommunications service providers undergoing deregulation in the United States and around the world. It is always useful to analyze the past and, especially, to enjoy the surprises—unlikely mergers, the rise of small carriers, and the uncharacteristically competitive energy of some former monopolists. Still, the more important concern is the future, namely uncovering which strategies are most likely to provide a path to differentiation in an industry that is still suffocating under its own commodity nature.

Paths to differentiation abound, and service providers will be wise to focus their investment and management energy on only a few. For many service providers that are accustomed to serving a universe of customers within a defined territory, focus itself will be a challenge. Service providers can select one or more paths to differentiate their services in four domains: planning, marketing, operations, and finance. Once the service provider has selected the strategic direction, it is essential to excel. *Strategies for Success in the New Telecommunications Marketplace* helps readers make strategic choices that are most likely to succeed, and it also provides tools for a hands-on resolution of the issues that arise.

No business book can hope to be completely current, even on its publication date. During the writing process for this book, mergers consummated, while others dissolved; price structures changed radically; and new technologies such as Internet protocol (IP) telephony moved from their formerly peripheral status to a new position in the center of nearly every business plan in the industry. Those facts that are committed to print but no longer valid should at least provide some insight into the strategic thinking of an earlier time and perhaps offer an enduring example even when the specific example is only history.

I would like to thank the coworkers and clients who have helped me learn to step back from the industry news to gain strategic perspective. In addition, I would like to express my appreciation to the manuscript reviewers, who each provided constructive feedback, and to the team at Artech House for its constant support during the development and production processes.

1

Map of the Market

That's an amazing invention, but who would ever want to use one of them?

—U.S. President Rutherford B. Hayes, after participating in a trial telephone conversation in 1876

Only a few decades ago, charting the components of the telecommunications market required a pair of entries: a fully integrated network for local service and long distance and the ubiquitous and featureless dial telephone. The forces of technology, globalization, and competition have shattered the landscape into many elements. Still, today's market more resembles the market of 20 years in the past—although dismantled—than a visualization of the market of 20 years in the future. Deregulation in most world markets has added competitors and reduced prices, but for the most part, the competitors are virtually interchangeable, and the array of services, aside from Internet access, is only slightly more diverse than it was in 1980. The dreams of real customer choice and an abundance of services still loom in the future.

The telephone handset itself [and its many variations, including wireless units, computers, network-enabled electronic organizers, automobile global positioning systems (GPSs), and so many other manifestations of hardware and software] has been uncoupled from the service provider's network. It is no longer central to a discussion of telecommunications services. In many communities, in fact, the telephone company offers customers Internet access and movies-on-demand but not telephones.

The proof of supplier interchangeability is the substantial *churn* (customer loss) experienced in the highly competitive markets of long distance, wireless service, and Internet access services. About 24% of households switched long-distance carriers in 1999, down slightly from 26% in 1998, according to research from the Yankee Group. A 1999 study from the Strategis Group found that 36% of wireless subscribers intended to switch carriers within the next year, and the likely churners represented the highest revenue segment of the customer base. Note that the wireless customers surveyed could not keep their telephone numbers from one carrier to another; if number portability were available, 42% expected to switch carriers. An earlier Strategis Group study showed that of the 10% of Internet service provider (ISP) subscribers who cancel service each month, two-thirds sign up with a different provider.

Still, in the early stages of deregulation, the monolithic monopolies have disbanded into a variety of service providers and supporting players. The first task of mapping the marketplace is to define the current group of performers. Note that many of the industry participants are still definable by their physical location on the network and the specific technology to which they are committed. Regulatory restrictions continue to limit location on the network and technology options for some providers, such as the regional Bell operating companies (RBOCs). By late 2000, the Federal Communications Commission (FCC) had authorized only Bell Atlantic (now Verizon) and SBC to offer long-distance services to their local customers, and only in one state each. More than four years after the Telecommunications Act of 1996, RBOCs have apparently failed to meet the requirements to lift the long-distance restriction. The choice of a technology platform for many of the industry leaders is simply an extension of earlier infrastructure, that is, circuit-switched networks. For other providers, such as some of today's fixed wireless operators and those companies committing to IP architecture, their locations on the network and the technologies they use are a matter of strategic choice.

Nonetheless, the slow pace of competitive development raises strategic questions: If selling wholesale services to local resellers is indeed critical to incumbent providers' long-range strategies, why do the RBOCs' resale customers protest that the wholesale offerings fail so significantly to live up to their needs? If serving consumers is within the short-range desires of interexchange carriers (IXCs), where are their competitive fiber-optic local networks? AT&T committed to a coaxial cable topology through its purchases of TCI and MediaOne. Why do the largest IXCs continue to build their local consumer networks almost exclusively through acquisition, rather than resale or buildout, both of which are likely to be lower cost alternatives? Competitive local exchange carriers (CLECs) and smaller long-distance carriers have taken the resale and buildout route instead. The simple explanation is that smaller carriers simply cannot afford the acquisition premiums. An alternative, somewhat counterintuitive, explanation is that the giants are not only well funded, but also willing to take the risks inherent in any acquisition. Perhaps the CLECs are willing to grow more patiently. Moreover, it is plausible that cultural factors simply prevent the interexchange behemoths from using strategic competitors as suppliers or installing high-cost, untried technology platforms. (Undoubtedly, no new entrant would build duplicate copper-wire local facilities.)

Incentives to take risks in the U.S. market have not been apparent to the incumbent industry leaders. RBOCs, eager to enter the long-distance market, appear to be willing to harvest their consumer retail businesses and invest in broadband network upgrades rather than develop a targeted wholesale strategy or risk failing in new territories. IXCs have been no more aggressive than were local incumbents in establishing facilities to serve consumers. The FCC requirement for SBC to offer services in 30 new markets—as a condition of the Ameritech acquisition—finally forced an incumbent carrier to take a business risk in uncharted terrain. When and if regulators approve more mergers between RBOCs, they will undoubtedly impose similar conditions. On the other hand, new entrants have innovated, and in many cases the incumbents have needed to become more creative in self-defense.

Figure 1.1 depicts the physical positioning of today's telecommunications providers on the telecommunications network. Access providers operate between the customer and the local switch or gateway into the network. Transport providers perform the function of transmission between local access switches. Services, software, and content can reside anywhere in the network. Their function is to enhance the telecommunications event with processing, information, or telecommunications support.

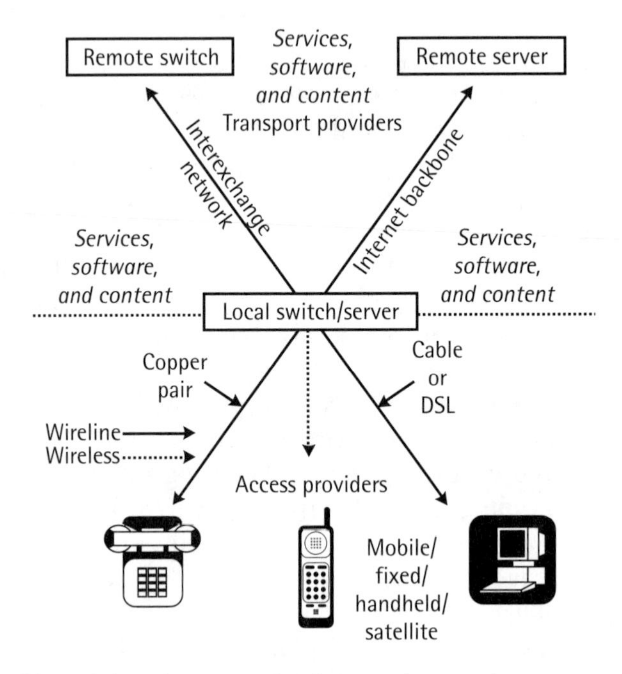

Figure 1.1 Map of the telecommunications services market.

1.1 Access providers

In the telecommunications network, the service between the first network switch or node and the user is described as *access*. Most access lines are twisted-pair copper wire, although new access technologies of fiber optic, coaxial cable for telephony and wireless access, are emerging. In most cases, access represents a fixed cost caused by a single user. Because the access facility is generally not sharable among multiple users, economies of scale are difficult to accomplish. Coaxial cable is sharable, which causes a new set of obstacles, as the service degrades when many users are connected. Rural users, far from the first network switch or node, cause above-average access costs, and represent the typical recipients of telecommunications service subsidies. Access is sometimes called "the last mile," referring to a normal distance between a consumer and the network switch. The United States lags several markets in South America and Europe in opening this market to competition.

Long-distance service providers pay access charges to distribute their calls in the local network. Access charges, priced above actual cost, constitute a large portion of the long-distance subsidies that support local service. RBOCs derive 22–27% of their revenues from access charges, and payment of access charges consumes about 40% of long-distance provider revenues [1]. Rural companies get more than two-thirds of their revenues from access charges and universal service subsidies [2]. In spite of its expense to telecommunications providers and the complexity of deregulating the market, access is the window to the customer and therefore a most strategically important position in the competitive marketplace.

Wireline

The wireline providers offer access to the network using a facility of copper wire, coaxial cable, fiber optic cable, or some combination. Except for coaxial cable, this facility is not sharable among users; it is either in use by the customer or waiting for the customer to use it. The access plant comprises a substantial portion of the local telecommunications provider's assets and overall investment, yet the access portion of a typical consumer's bill falls short of covering the provider's costs. Providers make up the shortfall through profitable services such as enhanced central office-based services. Examples of high-profit services include call waiting or voice mail, access charge revenues from IXCs for terminating long-distance calls, and narrowly defined long-distance services within a small territory. In the United States, non-Bell incumbent local providers are not restricted from offering long-distance services, but they often serve rural, high-cost territories, so it is still a challenge for them to earn back their access investment.

Several technologies enable the copper pair to carry broadband services. The first, integrated services digital network (ISDN), divides the copper line into three digital channels. Users can increase their Internet baud rate above those attainable on analog modems and make voice calls simultaneously. The maximum baud rate for ISDN is approximately 128 Kbps, which doubles the best rate available on a conventional modem. Newer digital technologies have already eclipsed this rate, and industry analysts currently view ISDN as a waning market for new investment. The newer copper-based broadband technology is digital subscriber line (DSL), which increases the baud rate to 256 Kbps and as high as 1.5 Mbps. Like ISDN, DSL enables the user to stay on-line while using the conventional voice network.

Companies offering voice and data service in this market include the RBOCs and the non-Bell local companies. The largest of the non-Bell companies, GTE and Sprint, were targets of acquisition or merger (with Bell Atlantic and MCI WorldCom, respectively). Hundreds of medium and small non-Bell companies continue to serve medium and small markets. New entrants to this market, CLECs, occasionally called integrated communications providers (ICPs), generally install fiber optic cabling to customers or resell the copper cabling of the incumbent carrier. Other CLECs, such as Covad, Rhythms NetConnections, Inc., and NorthPoint, have committed to installing DSL technology over resold incumbent copper. Because of their emphasis on data services, they most often call themselves data CLECs. Their business model is to upgrade the basic copper facilities of the incumbent carrier, representing access providers, such as ISPs, that sell the broadband service.

Perhaps because of the competitive pressure, DSL growth is substantial. Data CLECs are busily installing DSL broadband access on behalf of their customers, and incumbent local providers are rushing to bring their own services to market before the market is lost to their competitors. Pioneer Consulting produced the market estimates depicted in Figure 1.2.

CLECs, both voice and data, currently serve customers in more than 566 markets [3], with access line growth of more than 250% over the previous two years. While the first CLECs to market targeted the lucrative large metropolitan areas, market maturation has enticed CLECs to include smaller or less dense geographical locations, and they will undoubtedly be able to serve most consumers within the next few years.

Mobile wireless

This technology comprises cellular and personal communications services (PCS) networks. As new licenses have become available in major markets, competition has flourished. Wireless technologies continue their advancement. Costs have fallen. Consolidation in the wireless industry has created near-nationwide footprints in the United States for some carriers such as AT&T, Sprint, and global service mobilization (GSM) carrier VoiceStream. The resulting price structures among the competing carriers have enticed some customers to use wireless service nearly to the exclusion of wireline service. One estimate predicts that mobile wireless could draw half the market from incumbent providers by 2005 [4].

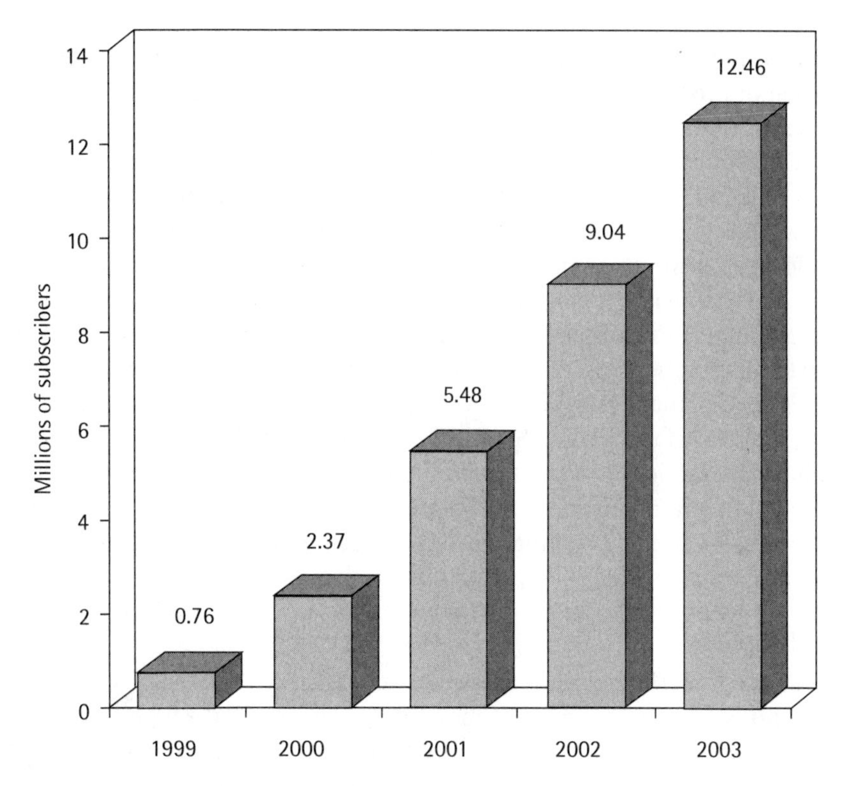

Figure 1.2 DSL projected subscriber growth.

The western European wireless market has progressed more rapidly than that of the United States. Cellular penetration has reached 50% in some markets and could soon exceed wireline density in some countries [5]. Between 1993 and 1997, mobile traffic in the United Kingdom grew by an average of 31% a year. Mobile phone penetration in Finland is approximately 60–70%.

Major U.S. wireless carriers include AT&T and RBOCs such as Verizon (Bell Atlantic and GTE), BellSouth, SBC, and AirTouch, now owned by Vodafone. In an effort to increase their networks' footprints, most of the largest wireless providers have merged with partners or entered into nationwide and sometimes global marketing agreements. Wireless provider Sprint experienced 145% growth in the nine months ending in March 1999, according to [6]. Omnipoint, a GSM carrier in the United States now merged with VoiceStream, demonstrated 88% growth in the same period. The continued growth in the wireless market has eased the burden of the high churn rate experienced by most wireless carriers. New buyers immediately replace customers who leave.

One source of growth in the wireless market is data access. Mobile workers and travelers are using their wireless telephones to collect e-mail and work on the Internet. Sherwood Research estimates that more than 32 million people in the United States spend 20% or more of their time away from their primary work environments [7]. New technologies promising higher baud rates, coupled with the affordability of wireless minutes, will help to foster this trend. The third generation (3G) of wireless technology is under development around the world. Asia is ahead of Europe, which, in turn, is ahead of the United States. General packet radio service (GPRS), a mobile data service, is in trials and will improve existing wireless data services until the more advanced 3G technology is ready for deployment. Furthermore, IMT-2000 standards will enable worldwide terrestrial and satellite-based networks to operate seamlessly.

Fixed wireless

The concept of fixed wireless access has generated considerable excitement in an industry characterized by labor-intensiveness and the high cost of installing the local access facilities. Indeed, the costs of the most basic wireless technologies are appealing, according to Allied Business Intelligence [8].

The greater bandwidth of wireline offsets its added cost, especially as the demand for fast data access rises. Nevertheless, the conventional wireless service as a substitute for wireline has proven to be valuable in areas that copper technology does not currently serve.

TABLE 1.1
Cost of Wireless and Wireline Technologies

	COST PER LINE (URBAN)	COST PER LINE (RURAL)
Conventional wireless	\$650–850	\$1,000–1,500
Local loop	—	—
Conventional wireline	\$800–1,000	Up to \$2,000

In mature markets, fixed broadband wireless technologies such as local multipoint distribution service (LMDS) and multipoint multichannel distribution system (MMDS) are taking hold. Both Sprint and MCI have invested in broadband fixed wireless; Sprint has made it a cornerstone of its integrated on-demand network (ION) offering. Local access providers Advanced Radio Telecom, WinStar, NextLink, and Teligent have made fixed wireless access integral to their business models. AT&T continues to experiment with local wireless with its Project Angel but has since changed its primary access model to cable, with its acquisitions of TCI and MediaOne. AT&T's current policy is to use wireless in areas not served by cable, depending upon its cost-effectiveness. The present challenges of fixed wireless technologies are technological and administrative. The technological challenge is that certain climates interfere with the signal (such as southern rainstorms). The administrative challenge is that standards for the technology are not established.

Handheld wireless

Wireless technology, generally for data access, is successfully in operation in handheld devices, especially telephones and personal organizers. Standards for wireless access protocol (WAP) and Project Bluetooth (the short-range radio standard for the 2.4-GHz band) have begun to launch a host of Internet-ready devices that use wireless protocols. The low bandwidth of small wireless devices probably means that their use for conventional Internet access will lag other access methods for some time. It is sufficient for short e-mail messaging and headline retrieval. While this is not an access technology in the most traditional sense, wireless always-on data access has significant implications for the remaining voice and data network. By diverting information to the handheld device, users will avoid inquiries by telephone or by Internet about the most ordinary requests, such as the status of a package, whether messages are outstanding, or the status of airplanes or train schedules. This technology thus has an inhibiting effect on more traditional communications venues.

Cable

While telecommunications service providers have always considered cable as a potential participant in the telecommunications marketplace, its role became more significant when AT&T committed to cable access for its consumer market. The announcement by America Online (AOL) of a merger with Time Warner enhanced cable's stature. Like copper, coaxial cable is a tested technology with a committed customer base. Nonetheless, it will require significant investment to perform well as a broadband access channel. Its major drawback is that customers share the cable data channels and that the connection rate degrades as many users concurrently use the shared facility. Moreover, the service and reliability reputation of cable service providers falls short of that of telephone companies in general. Customers will need incentives, in the form of low prices, high bandwidth, or quality of service assurances before they will switch to cable providers as their communications lifeline. Cable's primary advantage is that its existing infrastructure serves consumers, eliminating the need to build facilities to duplicate the copper network. AT&T's acquisitions of TCI and MediaOne offered it the opportunity to gain a facilities-based customer base and a large footprint in the consumer market.

Satellite

Satellite communication has hovered on the fringes of the communications market, but it could emerge as a significant alternative or complementary communications technology. Its most visible incarnations are broadcast satellite to the home using digital video broadcast technology and point-to-point satellite communications that boast a worldwide footprint. The broadcast architecture for currently deployed systems requires an uplink through the conventional telephone network, so it does not serve as a substitute technology for the present network. Geostationary satellites are high enough to mimic the Earth's orbit, so they appear stationary. They are inherently broadband, so they can compete successfully even as other wireless technologies improve. Their distance from the ground causes transmission delays (known as latency) that are familiar to viewers of televised international news events. Newer technology includes low Earth orbiting (LEO) satellites that circle the globe every one to two hours. One emerging provider, Teledesic, claims speeds that are comparable to fiber optic lines. Satellite systems excel in their reach, but so far, they have not been able to compete in price with more earthbound technologies. The ultimate market position of satellite technologies will most likely enhance other access mechanisms, rather than replacing them.

1.2 Transport providers

The term *transport* refers to the providers that carry telecommunications traffic from the originating local carrier to the terminating local carrier. In the United States and elsewhere, regulators found it convenient to separate access from transport as a means to deregulate part of the market without disrupting the other. The United States deregulated transport in the last two decades, and the resulting market has proven to be highly competitive. The importance of transport is that long-haul networks will be involved in virtually every telecommunications transaction in the future, from voice to data to video. If today's trends continue, bandwidth costs for transport will approach zero, and the network will move to commodity status. Alternatively, network enhancements can serve to differentiate transport providers competitively, creating new services and expanding the market.

Interexchange carriers

AT&T, MCI WorldCom, and Sprint are the best-known IXCs in the United States. Other IXCs that have chosen to build out a national network and to market the network to retail customers include Qwest, Broadwing, and Level 3. Prices for interexchange (also known as long-distance) service have dropped precipitously year-by-year. The market is quite elastic, and usage has climbed as prices have fallen. Advances in technology, enabling a fiber optic strand to carry more bandwidth, and the interest of well-funded providers have created a glut of capacity, further reducing prices. Fierce price competition and high churn characterize the interexchange market. Figure 1.3 depicts the market shares held by U.S. IXCs at the announcement of the now-failed MCI WorldCom/Sprint merger [9].

In the United States, customers select a preferred (presubscribed) IXC to simplify dialing. Unless the customer specifically dials around the presubscribed provider, the call travels on the designated network. This has created intense competition for subscriptions. In some European (until the European Union directive required preselection) and Latin American countries, such as Colombia or Chile, customers select the IXC on a call-by-call basis. The resulting nature of marketing and competition is widely different. U.S. carriers use telemarketing and direct mail to convert consumers so frequently that some customers receive several solicitations per week. In Colombia, where the basis of competition is different, each call is an opportunity to sell. Colombian provider ORBITEL has used raffles and

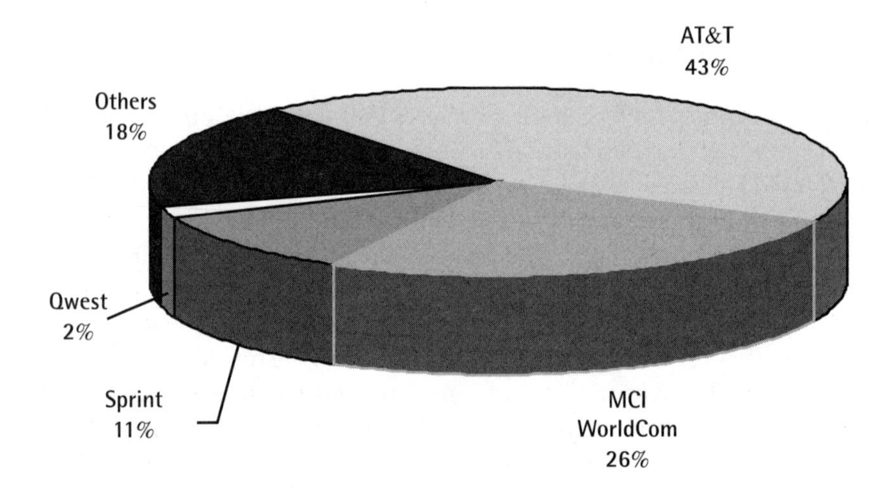

Others include:
Teleglobe 2.0%,
Williams Communications 1.8%,
Frontier 1.5%, Cable and Wireless 1.0%

Figure 1.3 U.S. long-distance market share.

prizes given away during the call to entice users to dial on its network rather than those of its competitors.

Interexchange resellers

Resellers of interexchange service purchase minutes in bulk from the facilities-based carriers and sell them at retail. Their rates are typically lower than those of the branded carriers, and several companies, notably Excel Communications, now owned by Teleglobe (in turn owned by BCE, parent company of Bell Canada), have become branded in their own right. According to market research group Atlantic-ACM, resale of long-distance services had a compound annual growth rate of nearly 20% in the United States between 1994 and 1997, about twice the rate of the total long-distance market. According to the Association for Communications Enterprises (formerly the Telecommunications Resellers Association), market share for the big three carriers (AT&T, MCI, and Sprint) was 81% in 1996, down from 90% five years earlier. Resellers accomplished a substantial part of that growth. In the United Kingdom, resellers report growth of 50–100%

per year [10]. The Yankee Group estimates that wireless resale will grow to more than $4.5 billion by 2002, a 7.7% market share of projected wireless industry revenues as compared to 4.4% in 1997.

Carrier's carriers

Some facilities-based providers choose to serve the trade rather than court end users. Williams Communications, once a pipeline company, built a network and served as a carrier's carrier before spinning off the telecommunications division. Enjoined from building new facilities for several years after the divestiture, Williams is again installing fiber optic cabling and again targeting only the carrier market. Metromedia Fiber Network (MFN) has built fiber optic rings in cities in the expectation that it would offer carrier's carrier services to local providers. In Europe, Unisource Carriers Service (UCS), a joint venture of KPN, Swisscom, and Telia, sells minutes of use on a wholesale basis to service providers. Wholesaler iaxis plans a pan-European wholesale network and has already built facilities passing 10 cities in Germany, Switzerland, and Italy.

Several other providers that have professed to be carrier's carriers have changed their business models. Global Crossing, once an avowed carrier's carrier with transatlantic cable facilities and fiber rings in various European cities, acquired retail customers with its purchase of Frontier Communications in the United States. Qwest Communications' early business plans were of a carrier's carrier, but its acquisition of US West places it among the top retail local service providers in the United States.

Internet protocol telephony providers

Historically, the analog telecommunications network operated on a circuit-switching architecture. Modems emerged to modulate and demodulate the signal so that digital data could traverse an analog network. Newer technologies remain digital from the desktop to the receiving desktop, in a packet-switched framework. Today's challenge is to carry voice transmissions successfully over digital networks, whether circuit-switched or packet-switched, such as IP, frame relay, or DSL.

IP telephony, as a replacement for the traditional circuit-switched model, is widely considered the future of wireline communication. The adoption of IP telephony through the Internet poses a short-term strategic and financial problem for local wireline providers, whose profitability depends on collecting high access charges from long-distance carriers.

Regulators have been reluctant to impose restrictions or subsidy payments on ISPs. Consequently, these companies are free to offer alternative long-distance services to customers at prices that are significantly lower than those on the traditional circuit-switched network. This interferes with the profitability of both the long-distance carriers and access providers. The quality of IP calls is still inferior to traditional calls but much closer in quality than it was in the past, and that problem is now only a short-term concern. The largest manufacturers have directed substantial investment at resolving the quality gap between analog and digital voice transmission.

Service providers, both local and interexchange, are quietly incorporating IP technology into their networks. This will probably create a regulatory dilemma when incumbent providers' networks are indistinguishable from those of ISPs, but the ISPs are presently free from the subsidies mandated of their interexchange competitors.

Internet service providers

Today's ISPs are only beginning to serve as telecommunications providers, but they represent a significant service provider in the near future. The merger announcement between AOL and Time Warner marked the first entry of an ISP into the local facilities-based access market, through Time Warner's extensive cable network. For long-distance service, ISPs offer IP technology to complete domestic and international calls. The prices for these calls, presently unencumbered by subsidies, result in a substantial discount for users who can tolerate the degradation in voice quality. For fast data rates, ISPs enter partnerships with companies to offer access through DSL and cable technology. Frequently, the local wireline provider is uninvolved in these upgrades. Thus, ISPs act as an agent to sell high-speed access over incumbent facilities through data CLECs.

ISPs are seeking new revenue sources, because profits from their fixed-access rates are less assured. As the Web becomes more user-friendly and sophisticated, users spend more time on-line, without providing new subscription fees. The highly price-competitive business ensures that rates will stabilize or drop rather than provide an increasing source of revenues. In the United Kingdom, AOL offers Internet access services with no subscription fee at all, for competitive parity with the locally emerging providers. Free Internet access is gaining ground elsewhere in Europe, South America, Asia, and the United States. In the United States, the growth rate for new subscribers is dropping precipitously [11], as shown in Figure 1.4.

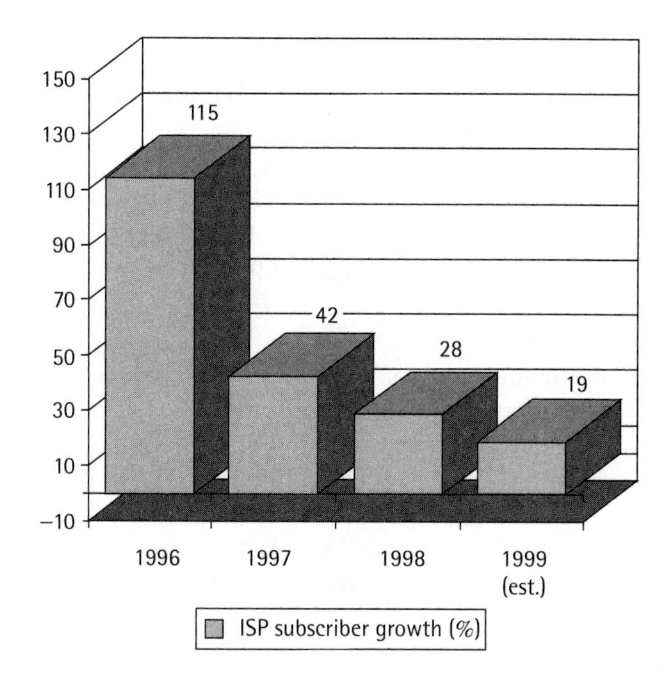

Figure 1.4 ISP subscriber growth is declining in the United States.

1.3 Services, software, and content

While most services, software, and content are beyond the scope of telecommunications providers, certain varieties are very strategic to the growth of providers today and in the future. These industry participants also represent a source of strategic competition to service providers, as providing telecommunications service is an obvious vertical integration opportunity. Firms in this sector have excellent access to funding, and their soaring market capitalization can make acquisitions easier to accomplish.

Operations support systems providers

Operations support systems (OSSs) comprise the software that manages the operations of the telecommunications network. They span the functions of connecting new customers; maintaining the network, billing, and customer care; and ensuring network reliability. Carriers spent $10 billion on OSS in 1999, and OSS expenditures are normally responsible for consuming 3% of carrier revenues [12]. The OSS market came about after long-distance deregulation. OSS developers include Clarify, Siebel, Vantive, and MetaSolv. Unlike monopolies, competitive providers will outsource functions if they are not core competencies, or not strategic and proprietary, or if outside developers can offer the function at a lower cost than in-house development. Competitive providers frequently outsource the OSS function, leaving incumbent providers with legacy in-house systems and merged systems that will require either an overhaul or a change in strategic direction.

Operations software has the potential to increase profitability for facilities-based providers and for wholesalers. For facilities-based providers, OSS offers the opportunity to enhance the basic reliability of the network and to differentiate commodity-like communications services through provisioning (setting up the connection), customer service and customer care, billing, and other enhancements not yet in place. For wholesalers, operations support will rise in visibility in situations where overcapacity characterizes the network from one end to another. Wholesalers will employ OSS to increase their own value as compared to their competitors. When a competitive retail and wholesale market is in place, retailers will expect wholesalers to develop core competencies in operations. OSS is the likely area of distinction.

Applications service providers

Applications service providers (ASPs) represent an emerging channel resulting from the widespread availability of the Internet. ASPs enable companies and individuals to use software on a rental basis rather than making an outright purchase. The fees for usage can be monthly, quarterly, or annual, or based on transactions, connect time, or other parameters. Part of the application and all of the data remain on the customer's local personal computer (PC), the remainder of the application resides on the ASP's servers. This will open the high-end software market to smaller and less affluent users. For example, small and mid-sized companies, trying to compete with industry giants, often do not have the in-house information technology expertise to develop and maintain systems, nor do they want to wait for a development cycle to go to market. They can still have access to the most sophisticated applications, such as the enterprise resource planning (ERP) systems that manage most business processes in an organization. Customers can pay for these applications on a usage basis, at a rate that is much more affordable than outright purchase from the software developer. There were approximately 100 ASPs at the end of 1999 [13]. ASPs include USinternetworking, Corio, and Telecomputing, which specialize in this market, and others, such as telecommunications service providers and software developers that recognize the business opportunity of rentable applications and the market's complementary position to their core businesses. International Data Corporation estimates that revenue in this sector, at about $150 million in 1999, will be $2 billion in 2003. A more aggressive estimate from Forrester Research predicted a growth rate for the ASP market of more than 400% from near zero in 1997 to $6.4 billion by 2001 [14]. Forrester later revised this estimate to a date of 2003 [15].

The industry is strategic to the telecommunications market for several reasons. Not only do rental applications add usage to the network, serving as an ASP also offers opportunities for telecommunications service providers and their partners to differentiate their services with network-based software offerings. Furthermore, small service providers that use ASPs as suppliers can increase their corporate capabilities at lower cost or lower risk than if they needed to purchase in-house systems.

Sun Microsystems moved its office productivity application StarOffice to the Web and made its usage free to gain market share. Database developer Oracle created Oracle Business Online to encourage its third-party developers to become ASPs. Microsoft has made overtures in this area as well. Operating as an ASP enables a software developer to reach out to the wider marketplace of small to medium businesses and individuals and offers a recurring revenue stream, always welcome between versions.

IXC Qwest and partner KPMG launched Qwest Cyber.Solutions to provide services to three defined market segments: applications management, applications hosting, and applications rental. Among its other motivations, the company views this area as an opportunity to develop strong customer ties while customers are small that will translate into service provider loyalty as the customer grows. US West, not normally considered a market trailblazer, has hosted Lotus Notes and Domino applications on a rental basis to its customers since 1996. It has since established a formal Business Alliance Partners program to maintain its lead in this market. An advantage of sustaining a server-based business for a company like US West is that most of the business is Web-reachable and does not require land or wireless connections over the company's costly to serve expansive and mountainous terrain.

Challenges exist in this market. First, customers need reliable broadband networks, either Web-based or over virtual private networks, to carry what are undoubtedly enormous applications with adequate response time. Second, pricing the rental fees too high will hurt ASPs' competitiveness against the outright purchase of software, notwithstanding the convenience of having an outside company run the application. Pricing applications too low will hurt the revenue stream for the software developer, which stands to lose sales to rentals from third parties. In addition, for roughly equivalent applications, so many providers have an interest in adding ASP services to their portfolios that price competition could become so fierce as to weaken profitability of the product line. Moreover, success in the ASP market will require partnerships characterized by tight integration and congruous objectives, not easy to achieve over the long term.

Web hosting is another form of ASP services, because customers depend on an outsider to develop their Web sites and manage the supporting software. ISPs and local service providers offering Internet access have found that Web-hosting revenues are valuable to augment their ever declining and ever flattening subscriber fees.

Portals

Portals are simply Web pages that are gateways or entrances to the Internet. They have proven to have value to investors and merchants, because so many Web users pass them in their on-line journeys. ISPs hope—or, in the cases of AOL and others, demand—that users stop at their own portals immediately upon logging on. With many users looking at the page on their way to searching or visiting, portals have become on-line malls and libraries, coaxing users to stay and search or shop or read news.

Portals such as Yahoo!, Excite, and AltaVista have a more difficult challenge, because their business model is primarily as a portal and not as an ISP, so they are unable to require users to visit as their home page. Several browsers allow users to choose any site they wish to be their home page; users of browsers that are independent of ISP home pages make up the target audience for portals.

While portals do not conform to the traditional model of revenue generation, by the end of 1999, the market capitalization of Yahoo! exceeded the capitalization of venerable service provider US West. The portal Excite has acquired Web-based greeting card portal Blue Mountain for approximately $800 million in cash and stock, notwithstanding the total lack of revenues of the acquisition target. This exemplifies both the market strength of portal sites and the importance of Web users, whether revenues are present or not.

For segmented consumer markets or vertical business markets, other portals are developing. On-line publications such as CNN in the consumer markets and trade publications in many business markets create portals for users to visit. These sites offer advertising revenues to the hosting company, and they offer carefully targeted audience members to advertisers. Portal sites have developed merchandising techniques to convince users to visit frequently, including free e-mail addresses or contest giveaways. Yahoo! offers an on-line calendar for its users to encourage users and their friends and colleagues to monitor the site regularly. The calendar, like its competitors', enables users to set reminders for tasks, create and manage events, synchronize to handheld devices, and reach calendar events from any Internet terminal. A portal-based on-line calendar offers an advantage to users over ISP-based calendars, because they can access their calendar from any ISP. Nonetheless, the ISP-based calendar benefits the ISP more than the user, because it serves as a switching cost for subscribers. For example, a user with a complex calendar on-line with AOL will have yet one more barrier to switching to a competitor. Unlike Web-based portals, ISPs will not forward personal information such as a calendar or an e-mail address to a competitor when the customer changes ISPs.

Portals are strategically important to telecommunications service providers. When the expected migration to IP architecture takes place, portals will serve to connect the access customer to its destination. Thus, portals represent a vertical integration opportunity for both access providers and transport providers. Conversely, communications services will be a vertical integration opportunity for well-funded branded portals.

1.4 Convergence

It is necessary at this point to emphasize that the market participants by no means willingly inhabit their niches passively. First, some of the largest access providers are eagerly awaiting the elimination of restrictions against entering interexchange markets. Carriers not currently serving local customers are building out their networks as quickly as capital becomes available. Providers that view their current revenue streams or business models as limiting are seeking ways to increase revenues or change their timing, reduce dependency on less profitable lines of business, or position themselves for future markets.

Part of the challenge facing all of the market players is that they want to broaden their reach, but there is only room for a limited number of giants. They want to maintain their share in existing markets, but competitors are rushing in from known directions and from some unexpected directions as well. The market's pace will force managers to bet the company routinely on anticipated but uncertain market outcomes.

The remainder of *Strategies for Success in the New Telecommunications Marketplace* will analyze which strategies are most likely to succeed and which are less desirable for contenders in this fiercely competitive arena.

References

[1] Booz-Allen & Hamilton, "The Last-Mile Dilemma: Strategies in the Local Loop," *Insights*, Vol. 3, Iss. 1.

[2] Lawyer, G., "Rural Retreat," *tele.com*, Vol. 3, No. 6.

[3] Barnich, T. L., "State of the CLECs: 1999," *America's Network*, Vol. 103, No. 11, p. 14.

[4] Hill, G. C., "Telecommunications (A Special Report): Bypassing the Bells—Consultant's Call," *Wall Street Journal*, Eastern edition, Vol. CCXXXIII, p. R27.

[5] Rocks, D., "Western Europe's Mobile Onslaught," *tele.com*, Vol. 3, No. 9.

[6] Harter, B., "Wireless Review—Top 25 Wireless Carriers," *Wireless Review*, Vol. 16, No. 17, p. 14.

[7] Riner, J., "Nudging Data Forward," *Wireless Review*, Vol. 16, No. 9, p. 38.

[8] Mentrup, L., "Pulling Out All of the POTS," *Wireless Review*, Vol. 16, No. 21, pp. 64–68.

[9] "MCI, Sprint Could Pass Antitrust Test," *Wall Street Journal*, Eastern edition, Vol. CCXXXIV, No. 61, p. B6.

[10] Finnie, G., "From Resale to Retail," *tele.com*, Vol. 2, No. 3, pp. 47–48.

[11] Mine, H., "Internet Globalization: Here and Now," *tele.com*, Vol. 4, No. 19.

[12] Williams, W., and P. B. Robinson, "What's It Worth?" *Telephony*, Vol. 237, No. 3, p. 36.

[13] Quinton, B., "Kicking ASP," *Telephony*, Vol. 237, No. 18, pp. 38–54.

[14] Mitsock, M., "Telcos Find Home in ASP World," *Phone+*, Vol. 13, No. 13, pp. 142, 144–145.

[15] Gerwig, K., "Reality Check," *tele.com*, Vol. 4, No. 16.

2

Today's Telecommunications Landscape

A 1999 Bloomberg report said that a Zimbabwe father rejected the traditional marriage price of cattle in favor of a wireless telephone.

It is not surprising to see the rush of new entrants in the global telecommunications marketplace, the expansion plans of existing carriers, or the merger and acquisition frenzy all over the world. The telecommunications market is immense and growing at an enviable pace. Figure 2.1 depicts the mix of revenues in the telecommunications industry [1].

According to Probe Research, the global market already represents about $800 billion in annual revenues. A 1999 report published by the U.S. Council of Economic Advisors estimated that the industry created 200,000 U.S. jobs in the last decade. The Telecommunications Industry Association found that transport services in the United States grew at 8.5% in 1999 to $252 billion and that support services increased by 17.3% to $138 billion. The U.S. long-distance industry grew at an annual rate compounded to 7% from 1984 to 1999, according to research firm Atlantic-ACM. Mobile services continue to grow at double-digit rates, but the most impressive growth

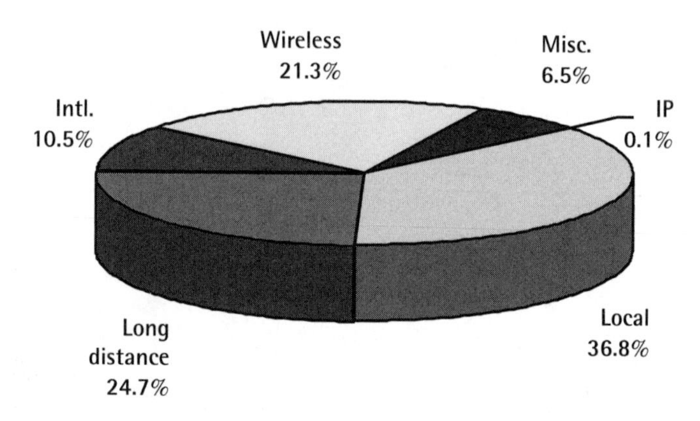

Figure 2.1 Global telecommunications revenues (1999).

is currently in data services, estimated at a rate of 30–80% by a variety of sources.

2.1 Unique market characteristics and challenges

Like all markets, the provision of telecommunications services conforms to the basic laws of competition. Still, certain anomalies in the market are notable.

New services require an education process before the sales process can take place. Telecommunications service providers sometimes offer less expensive or more efficient alternatives for customers, such as digital wireless rather than traditional cellular service. These are relatively easy to explain and sell to customers. Other examples, at least for customers already online, are DSL and ISDN, which enable customers to connect at a higher baud rate to the Internet and allow simultaneous voice conversations.

Nevertheless, it is necessary to explain the benefits of truly new features before the selling process can take place. Examples include central office-based services such as call waiting, call forwarding, and other enhancements. Market research to design pricing models historically produced flawed results, because customers had no idea of their willingness to pay for a service they barely imagined needing. Typically, RBOC monopolies would conduct a survey to assess customer interest in new services. The surveys often concluded that the potential market was small, in which case the price would need to be relatively high per customer, or that the market simply did not justify deployment of the new technology. As enhanced central office services required only a small incremental investment in the switch, companies sometimes offered the services anyway and discovered that customer interest far exceeded the survey predictions. Promotions that include free usage for a short period are very effective for new services. This is one rationale behind the scores of introductory disks and preinstalled software offered by ISPs to new users. New networks, especially IP networks, will be more feature-rich than today's infrastructure, so this challenge will rise in importance in the coming decades.

The rapid pace of technology advancement is arguably the primary market driver, yet most telecommunications service providers do not directly affect its direction. Most telecommunications service providers do not presently support an in-house research and development (R&D) function. Companies are indeed ensuring that their networks are state-of-the-art and that the quality of service is first-rate. The focus simply has not been to build a proprietary excellence.

The technology infrastructure is less direct in the telecommunications services market than in, for example, the PC market. Undoubtedly a major reason is that a services business competes on many variables, only one of which is the quality of its network hardware. Equally important variables are the management and operation of the technology. The fact that facilities-based providers have not actively invested in proprietary technology is most likely a temporary situation. This might have resulted from clumsy organizational structures in an immature marketplace, commodity status of most services, and the lack of defined business models in a gigantic competitive arena.

Some companies do not support in-house development facilities because of legal restrictions, or because the market for their technologies is larger if it is shared outside the walls of the organization. Bell local companies face a manufacturing restriction as a condition of their divestiture from AT&T, and this limitation, coupled with an abundance of alternative investment opportunities, has discouraged direct technological innovation. The remaining vestige of their R&D capabilities was their joint ownership in Bellcore, which its original owners, primarily the RBOCs, sold to Science Applications International. AT&T, formerly a leader in R&D and manufacturing, divested the manufacturing entity Lucent for reasons of channel conflict, in which it dismayed its customers by competing against them. A leader in technology, Nortel is a primary equipment provider for its sister company Bell Canada, although channel conflict is not presently a problem for it outside of Canada, where the potential market is quite large.

Whether providers will take more control over their own technology destinies will depend on several factors. The Bell companies, accustomed to delegating the pace of technology deployment to their former parent organization AT&T, required a cultural change to demand control of their own infrastructure, secondary to other necessary changes to the marketplace mind-set. Some telecommunications providers with a more competitive approach will decide outright that technology leadership is or is not a part of their business model. As telecommunications services remain a commodity in the early stages of competition, and standards are not universal among emerging technologies and platforms, few companies are yet willing to make large investments in manufacturing and research.

Market information is not available for new products. Telecommunications services, especially those supporting or supported by personal computing, often change the way users live or conduct business. Early computer-based technologies such as answering devices, spreadsheets, or word processors simply made existing activities more efficient or less labor-intensive. They automated activities rather than creating new activities. Some highly innovative computing still automates what people already do rather than creating entire new environments. Web shopping sites for branded items such as booksellers, electronics stores, and even clothing retailers offering virtual dressing rooms simply eliminate the travel to a store and do not radically change the way the shopper lives. Other communications and computing activities such as call waiting and on-line auctions have few manual equivalents. Because the new technology does not replace an existing cost, deriving the benefit to customers, and accordingly the price, is close to speculative. Moreover, since the incremental costs of adding functionality to existing systems and networks is often at worst a fixed cost and at best a small incremental cost, pricing services for profitability can be challenging.

Timing the deployment of services, within strict limits, is essential. Time to market is irrelevant in most monopolies, even technology-based ones, where cost recovery for installed equipment generally can occur over a long product life because more sophisticated competitive offerings are not available. Time to market is a success factor in competitive technology businesses. When Intel delays a chip by weeks or months, its stock shudders. When monopoly-based cellular service took a regulatory lag of decades between invention and deployment, few stakeholders noticed.

Companies moving from regulation to competition will learn to launch products quickly.

In a competitive market, not only are competitors everywhere, but information about their own deployment schedules is only as good as competitive intelligence. There is substantial pressure to move to market as quickly as possible.

On the other hand, customers are less tolerant of network failure than they are presently of market delays. A telecommunications service provider needs to have the infrastructure (technology and business processes) in place to provide equivalent service to existing offerings. If the provider deploys the service without the supporting infrastructure, the new service could fail on its own or harm other services residing on the same network, and the provider risks damaging existing brands. It is a difficult tradeoff, and the quantity and quality of market information available to the provider can be the difference between success and failure.

Revenue growth for an individual provider can come from a variety of sources, making it more complex to calculate one's market position. All companies follow their own revenue and income growth carefully, but knowing the corresponding data for competitors is not an easy task. While public companies do announce gross revenues and profits, an individual competitor's financial data needs adjustment to eliminate unrelated divisions' revenue sources, territories not similarly served, and other artifacts of the analysis. Determining growth is one challenge, while verifying one's own market share is quite another.

While revenue growth is always a goal, revenue growth by itself does not validate a company's strategies, marketing, or operational soundness. Most telecommunications markets do benefit through higher volumes from falling prices (due to regulatory changes and network costs) and the expansion of the Internet user market. The wireless user base continues to grow as wireless services become more functional, more affordable, and more mainstream. Businesses and consumers routinely buy duplicate facilities for telecommuting, on-line access, fax, and wireless connectivity. Other than the most established competitive markets, it is hard to find a telecommunications services niche that is not experiencing double-digit growth. If a company is experiencing double-digit growth, it is difficult to ascertain whether it has become more competitive or if it is simply riding a market wave.

The fact that prices for comparable services are falling mitigates the explosive growth. Companies do have to sell more than the previous year to improve gross profit, even when costs fall in proportion to prices. Local telecommunications service prices are rising in a restructuring to better reflect cost. Besides that, the only prices in the sector that are rising are in cable television (CATV) subscriptions since the unconditional relaxing of pricing restrictions in the Telecommunications Act of 1996. Therefore, price decreases add complexity to the equation and cancel out some of the higher revenues that simpler growth would provide.

Churn, the turnover of telecommunications customers, changes one participant's view of market growth with a corresponding and opposite change to another's. According to the Yankee Group, carrier-to-carrier churn is higher among high-volume users, at about 33% among those who use more than 200 minutes per month. Churn rates will remain high until markets stabilize, until competitors operate on an equivalent basis, or until telecommunications service providers improve their ability to differentiate their offerings from those of their competitors. An unusually effective customer acquisition campaign by one competitor can reassign the customer base temporarily, and at high churn rates, market share estimates can appear more robust or gloomier than they really are.

Many industry participants are either new to competition or new to telecommunications. The clash between incumbent facilities owners and competitive resellers is not only about competition. There is a cultural divide between the largely entrepreneurial new entrant's mind-set and the traditional protectiveness of the facilities-owning incumbents. Notwithstanding any actual obstructionism that might take place, entrepreneurs prioritize time to market with a confidence that they will handle operational problems as they occur, and incumbents traditionally favor conscientiousness over responsiveness.

Those that are torn between both mind-sets, such as AT&T, find themselves on both sides of competitive clashes. One likely reason that AT&T committed to a coaxial cable technology as its entry strategy into consumer local access markets is its recognition that incumbent local companies would be protective of their customer base and their facilities for resale. Still, when regulators expected AT&T to resell its cable access to competitors, its adamant opposition appeared to be the same position it had criticized when it came from local copper-based incumbents. This cultural collision will diminish and eventually disappear as the market matures and the cultural differences diminish between incumbents and new entrants. Other world giants, such as Deutsche Telekom and France Télécom, have proven to be similarly aggressive and entrepreneurial in their foreign ventures but still somewhat protective within their home markets.

The optimal size for an industry leader is unknown, but it appears to be getting larger. AT&T divested seven operating companies at roughly the same size, and significantly below its own remaining size when measured in revenues or assets. AT&T retained $34 billion of its approximately $150 billion of assets, with the remainder divided approximately evenly among the seven RBOCs. None was large enough to confront AT&T directly, even if market conditions were favorable for that, and they decidedly were not favorable; they are barely adequate nearly 20 years later.

Mergers and acquisitions have characterized the U.S. and European markets, demonstrating that providers believe that critical mass for success is larger than their present structures. Bell Atlantic and NYNEX merged, and GTE has joined them. SBC acquired Pacific Telesis, SNET, and Ameritech. WorldCom bought many companies before it purchased MCI and announced that it would incorporate Sprint into its corporate umbrella. AT&T made major acquisitions in the cable industry. In Europe, acquisitions by Mannesmann and Deutsche Telekom, a merger between Vodafone and Pacific Telesis' wireless spin-off AirTouch, a long-awaited merger between Vodafone and Mannesmann, and new mergers announced regularly all intend to create huge companies with global aspirations. On the other hand, a merger attempt between Sweden's Telia and Norway's Telenor dissolved during the planning phase. A similar derailment occurred when KPN and Spain's Telefónica de España intended to merge and become the fourth largest telecommunications service provider in Europe.

Other than AT&T's venture into cable access, the commonality of recent mergers is that companies tend to join with others in the same business. AT&T's experience with computer equipment manufacturer NCR, Bell Atlantic's failed merger attempt with TCI, and IBM's Internet and telephone equipment experiences have made telecommunications providers uneasy about straying too far from businesses they know.

International service providers have added partnerships to their merger declarations. AT&T and British Telecom made a significant organizational commitment to their Concert partnership. Qwest and Netherlands provider KPN are in a partnership to create a pan-European network. Global One, a consortium comprising Deutsche Telekom, France Télécom, and Sprint, was threatened by Sprint's acquisition by WorldCom; the two remaining partners purchased Sprint's shares and eventually the partnership dissolved entirely, as did the Sprint/WorldCom merger.

These initiatives could represent business judgment based on sound analytical frameworks. Then again, they could be a diversion to management attention and offer a rationale to procrastinate from real competitive progress. Alternatively, growing through mergers might be the path of least resistance for senior managers who do not have the energy to grow as fast as younger, leaner competitors the old-fashioned way, by building gradually. Last, relative size could matter more than actual size. Competitive jockeying for position can be a survival strategy instead of an ego gratification. If this is true, service providers making what appears to be a rational decision to stay small could decline and themselves become targets. Eventually, history will record some mergers as winners, and others as errors.

2.2 Role of the Internet

The Internet's rise has transformed many aspects of commercial operations in virtually every industry, from advertising, sales, and customer care, to telecommuting opportunities for employees, to employee communications. For telecommunications providers, the pervasiveness of the Internet has also changed business volume, distribution, and network architecture.

Web usage (or its variations) contributes significantly to the growth in access lines and communications activity of recent years, and will be responsible for most of the anticipated growth in the future. Not all growth is due to Internet usage. Dropping prices due to regulatory changes, increased competitive alternatives, productivity gains, and advances in technology create new demand in elastic telecommunications services. Still, Internet growth has resulted in second-line purchases and a substantial base of ISP subscribers. Furthermore, customers already using the Internet are seeking faster response times through higher bandwidth, resulting in growth for highly profitable broadband data services. While growth in consumer subscriptions to the Internet has slowed, new users in small business have strengthened the market.

Like companies in any other industry, the Internet supplies a new distribution channel. The customer segment that uses on-line distribution today is small but demographically attractive, comprising customers who buy more services and produce high usage. Incumbent telecommunications providers have not been leaders in Web site launch or for creativity once in the market, although new entrants are using on-line distribution to reduce costs and differentiate their services by creating a progressive brand. Telecommunications service providers can offer services on-line; computer makers and booksellers have developed on-line shopping for years. On-line commerce should be a leadership opportunity for telecommunications service providers, but to date their performance has been ordinary. The most creative uses of the Web in the telecommunications industry are the sites that assist customers in making telecommunications service purchases, sites built by entrepreneurs, not industry insiders. Nonetheless, telecommunications service providers have made inroads into on-line self-service and provisioning, bill review and payment, service upgrades, trouble reporting, telephone listings, electronic bonding (the exchange of information between carriers), and other Internet-based business opportunities. The on-line channel reduces costs and improves the customer experience for those who prefer independence and 24-hour availability.

The Internet is driving the transformation of the network from a circuit-switched to a packet-switched IP architecture. It has produced the broadband "killer application," awaited for a decade by telecommunications service providers. While voice still commands most network traffic, virtually every telecommunications service provider is preparing to invest in an IP architecture for the long term. New interexchange providers such as Qwest and Level 3 have made a strategic commitment to IP. All of the traditional carriers have IP programs in place, and the major switch manufacturers are manufacturing switches and other equipment to manage the migration from one platform to another.

From a regulatory standpoint, the Internet represents solutions and new challenges. Two decades ago, in an effort to maintain universal service without restricting the advancement of computer technology, the FCC erected barriers between the computing and communications markets. While this approach provided temporary guidance, the concurrent development of the sectors made the distinction impossible to sustain. The FCC continues its resolute opposition to regulating Internet-based activities, including voice communications over the platform. Today's IP telephony services are available outside of the access charge structure imposed on traditional interexchange communication. The continuation of subsidies only on conventional circuit-switched networks creates an arbitrage opportunity for Internet-based communications.

2.3 How mergers have changed the market

Telecommunications mergers, which have dominated industry news, have also been highly visible in the general business press. According to Broadview International, the 10 largest mergers in the sector accounted for 40% of the total value of all technology transactions globally in 1998 [2]. In Europe, telecommunications mergers and acquisitions accounted for 30% of global technology activity during 1999, with the value of deals increasing more than 22 times over 1998.

Whether or not they provide profit to their own shareholders, mergers have changed the market in direct and indirect ways. Mergers boost both the market's and management's perceptions of the optimal—and the minimal—size for a telecommunications service provider. Therefore, mergers tend to create more mergers. A company that achieves a national footprint, whether through growth or mergers, will have opportunities that are unavailable to companies without the same resources. Nationwide companies can advertise nationally, which is more cost-effective than selecting markets and advertising within each one. Companies with nationwide or global facilities can serve the highest-volume and most profitable business customers.

A company offering national or global services has less interconnection and revenue sharing with other providers. Broad coverage can offer pricing alternatives that are unavailable to one's competitors. AT&T revolutionized wireless service with its Digital One Rate nationwide wireless service, as Chapter 3 describes in depth. GSM provider Omnipoint offers global roaming, and Vodafone has announced its plans to offer the same capabilities through its satellite network and its association with land-based partners.

Regulators have become more experienced at merger approval reviews, and in the process, they have become more creative and demanding. Most approvals of U.S. mergers included FCC conditions to adhere to the original 14-point checklist outlined in the Telecommunications Act of 1996. The checklist identifies operating procedures and other practices that will facilitate competition, such as unbundling network elements, providing adequate access to facilities and services, number portability, dialing parity, and access to needed information. Later merger approvals required aggressive discounts in a company's own territory to resellers of its services and a requirement to serve less desirable markets with broadband services. The conditions imposed on SBC added a requirement to become a CLEC in 30 markets outside its 13-state territory by early 2002, with hefty penalties for noncompliance. The FCC recognized that incumbent RBOCs had been reluctant to expand in local service outside their existing territories, and used the Ameritech merger opportunity to boost facilities-based competition. New applicants for FCC approval can expect similar conditions for their own merger requests.

2.4 Non-U.S. markets

Europe

Europe, the world's second largest economy, accounts for one-third of the $1 trillion global telecommunications market for services and equipment [3]. Backbone providers are installing bandwidth, and telecommunications service providers are building fiber optic rings around metropolitan areas. Companies involved in building new networks include European incumbents and newer service providers such as Global Crossing, Hermes Europe RailTel (now called GTS), MCI WorldCom, Viatel, and Level 3 International.

Europe officially declared deregulation with the EC Directive for Liberalization in 1998, although deregulation was already under way in the Nordic countries and the United Kingdom. Only five countries—Austria, Denmark, Finland, Germany, and the Netherlands—had unbundled their local loop infrastructures to make them available to competitors after the first quarter of 2000.

Competition within Europe has characteristics that differ from competition in the United States. First, the presence of distinct incumbent carriers in each country creates a more competitive landscape for each company's expansion. Wireless has a more significant presence in European countries than in the United States, which will make fixed wireless technologies easier to incorporate into access alternatives for customers. Carrier preselection, the ability of a customer to presubscribe to the long-distance carrier of choice, a fixture of U.S. competition since the mid-1980s, was not a requirement until 2000 in European markets. The same 1998 directive required number portability, the ability of a customer to change providers without changing telephone numbers, a condition that is consistent with the FCC's checklist for RBOC entry into long distance. Growth is about 30% higher than in the United States, but churn is higher as well, at about 35–50% annually, and average revenue is declining at about 15% a year [4]. In the first 18 months after deregulation, long-distance and Internet service providers in the most open markets dropped their rates by about 50% [5] and by as much as 70% in Germany.

The cable infrastructure in Europe is newer than that of the United States, but penetration rates are somewhat lower. Cable penetration for television is higher in Northern Europe than in southern countries. Cable operators are optimistic that multimedia applications such as video on demand and broadband data access will encourage more users to subscribe. Hindering cable as a competitive access alternative is the fact that in many countries with the highest cable penetration, the local telephone company is the cable provider.

Internet access is not as mature in Europe as it is in the United States. Many providers have a regional, or at most, a national reach. As of mid-1999, PC penetration in Europe was at about 20%, or about half that of the United States [6]. Enterprising ISPs have tried to overcome the usage-based access pricing that is prevalent in Europe by offering free Internet access. AOL needed to offer free service, unprecedented for the U.S. market leader, in the United Kingdom to meet competitor packages. Recent initiatives in the United Kingdom and elsewhere have removed the secondary barrier of metered usage. This will significantly reduce the penetration gap.

The United Kingdom is among the most competitive areas of Europe, with an above-average-sized Internet sector and more cable penetration than other countries, owing to early deregulatory efforts over the last 15 years. Nevertheless, most residential customers, especially those outside the large cities, do not have a choice of fixed-line providers [7]. Former monopoly provider British Telecom still controls nearly 85% of the local loop and will lose its monopoly status by July 2001. Regulatory authority Oftel has proposed to roll back its regulation in favor of more self-regulation by the industry, with more than 200 service providers in competition. British Telecom faces competition in local service from competitors from the United Kingdom, elsewhere in Europe, and the United States. New ISPs offering free subscriptions induced British Telecom to eliminate the metering of on-line connections, which will undoubtedly encourage additional usage.

Prices for long-distance service dropped by 40% in a two-year period in France [8]. Still, the long-distance market share loss for incumbent France Télécom was not as significant as the experiences of other countries undergoing competition. Nonetheless, to offset the revenue loss, the company launched a rate rebalancing and more than doubled local loop charges. The government, a major shareowner of France Télécom, has not opened the local infrastructure to competition, but competition in the local loop could take place as early as 2000. France Télécom has focused its investment on Europe, purchasing 10% of U.K. cable operator NTL Group and a majority interest in German wireless provider E-Plus Mobilfunk GmbH. Furthermore, the service provider plans to build a European backbone network that will connect 40 European cities by 2002.

Deutsche Telekom, the incumbent carrier in Germany, has lost customers in the high-profit business market. Numerous studies demonstrate that a sizable number of business customers either have switched to alternative carriers or are seriously considering doing so with some or all of their service [9]. According to the Yankee Group, new carriers have taken about 35% of the long-distance calling market from Deutsche Telekom. The country's strategic location makes it an attractive hub for competitors wishing to serve all of Europe.

Switzerland is undergoing significant competitive entry, and its market sustains active price competition and increasing mobile penetration. Sunrise and mobile carrier diAx base their service infrastructures on the existing fiber networks of rail and utility companies. The local loop is still not competitive, but large cable operator CableCom represents a significant strategic competitor. The mobile market is responding to competitive entry. diAx offers location-based wireless information services to customers using GPS data.

Regulators in the Netherlands have aggressively licensed competitors in the interest of fostering competition. CLECs such as MFN are building facilities in the local access network. Foreign companies, including MCI WorldCom and British Telecom, have established facilities in the Netherlands, and incumbent provider KPN has invested aggressively in Europe and elsewhere, most notably in partnership with U.S. provider Qwest. The effect of competition is most apparent in the business market, where at least one-quarter of companies use a service provider other than incumbent KPN.

In Belgium, long-distance market share loss may only be approaching 5% [10]. Most of Belgium's residential customers have a choice of fixed-line local service providers, due to the market entry of Mobistar, a mobile provider. Moreover, competitive cable providers are present in nearly all households, and the CATV industry is actively upgrading its infrastructure. Network providers such as Global Crossing and GTS have included Belgium in their fiber infrastructures.

Scandinavia is in the lead in fostering and achieving competition and in customer acceptance of new services. Finland deregulated local and long-distance service in January 1994 and international calling in July 1997. Finland boasts a mobile penetration approaching 70% [11], attributed to initiatives such as calling party pays (CPP), separating handset costs from usage charges, and implementing standards making it easier for customers to switch providers. Sweden's market was never a monopoly, but no carrier competed with its incumbent Telia until 1993, when regulators developed a framework for competition, revised in 1997. With five mobile networks and a mature cable infrastructure, Sweden has been a leader in offering a business opportunity for competitors. Denmark liberalized its telecommunications sector in July 1996, and new entrants, foreign and local, have become telecommunications service providers. In July 1999, Norway effectively eliminated the distinction between local and long distance by establishing one rate zone for the entire country. The country has a high ISDN penetration rate, which incumbent provider Telenor attributes to low pricing and easy self-service installation.

Competition in Italy's telecommunications sector is influenced by the acquisition of the incumbent provider Telecom Italia by Olivetti and the introduction of strong competitor Mannesmann, after its purchase of majority interests in Infostrada and Omnitel. Italy represents Europe's largest cellular market; mobile users surpassed fixed lines in 1999. Liberalization was relatively recent, and the local loop is not yet competitive, nor is number portability available. Moreover, dialing a long carrier code and the unavailability of carrier preselection have impeded long-distance competition. Without a cable infrastructure, mobile presently represents the best alternative access technology.

Spain's local market has been relaxed by using indirect services that rely on the resale of incumbent Telefónica's local facilities. Still, competition from new entrants has created downward price pressure and has offered an entry strategy for providers to gain customer share before building their own facilities. Compared to other countries, competition is less active, but the introduction of facilities-based carriers will allow preselection and create a more dynamic market. Telefónica has fiercely defended its incumbent market position but has aggressively invested outside of

Spain, primarily in Latin America. Also, its Internet subsidiary, Terra Networks, acquired Lycos, a leading U.S.-based portal and search engine.

The Americas

Within a regulated environment, Canada had a system of regional telecommunications providers, some owned by government. In 1985, Canadian regulators opened competition in the long-distance market, first through resale, then through full-fledged competition seven years later. KPMG estimated that rates were 55% lower than they would have been if competition were not present and that usage had grown by 67% [12]. The country is now poised to open the local service market to facilities-based competition. Well-funded competitors include Bell Canada, Bell Nexxia, AT&T Canada, Sprint Canada, Aliant (made up of the four eastern Canadian incumbent local providers), Intrigna, Telus, and Teleglobe (now part of BCE).

The level of competition and deregulation in Latin America varies from among the most deregulated environments to the continuation of highly regulated services. The economics of the region and the preponderance of sparsely populated and underserved areas create both a challenge and an opportunity for providers to serve the region. Foreign giants such as Telefónica de España and BellSouth have made express overtures when opportunities arise.

South America's telecommunications service penetration rate is approximately 10%. The quality of Latin American infrastructure is inconsistent, and some landline networks do not lend themselves well to upgrades. This results in an opportunity for fixed wireless alternatives for local service.

Chile privatized and liberalized its telecommunications sector about a decade ago. Because Chile is recognized among the most competitive environments in the world, carriers and overseers learned lessons about competition that will be useful to newly competitive markets. In Latin America, Chile was first to liberalize long distance, cellular and PCS, and cable telephony. Chile boasted the first public IP telephony and fax services and the first unbundled local carrier network. When the country privatized long distance, nine competitors rushed in. The downward pressure on prices was so great and margins were so thin that service suffered. Eventually only three of the carriers were profitable. Carriers are now cooperating to establish a unified billing system. Competition has driven providers to install fiber optic networks and sophisticated switches.

Brazil privatized its incumbent telecommunications provider and awarded wireless licenses to encourage competition. It established three regional wireline companies, an international long-distance carrier, and awarded licenses to nine wireless spin-offs to compete with these service providers and each other. Its economic crisis has interfered with the growth of the market, but progress is nonetheless evident. Forecasts for its wireline market are robust, and wireless forecasts are even brighter. The wireline market is expected to double from mid-1998 to 2003 [13]. From 1994 to 1998, its wireless market grew tenfold, and Frost and Sullivan estimated that it would nearly double again by 2000. Internet estimates are even more aggressive.

Mexico privatized Teléfonos de México (Telmex) in 1990, and the two largest of the 10 competitors have captured 10% of the long-distance market [14]. According to CIT Publications, alternative operators had taken 25% of the long-distance market from Telmex by the end of 1998 and 40% by the end of 1999. Mexico has also completed its wireless PCS auctions. As local service has been slower to move to competition, wireless service, both mobile and fixed, acts as an alternative to the incumbent carrier's service. Cellular penetration increased by 116% between 1998 and 1999. Mexican regulators have managed the transition to competition vigilantly and have adjusted market conditions or pricing structures to ensure a smooth progression.

Argentina determined that it would eliminate monopolies held by Telefónica and Telecom Argentina. Companies such as Diginet Americas and Formus are building networks in the region. Unlike some other South American countries, foreign investors are not reluctant to finance programs in Argentina. U.S. providers BellSouth and WinStar are building facilities in the region. Regulators increased the duopoly held by mobile carriers to three-operator structures, anticipated to create additional competition.

Asia/Australia

Asian countries are also undergoing deregulation under the control of their regulatory authorities. Access to new technologies is encouraging the growth of communications, but Asia's overcrowded cities and dense infrastructure present their own challenges.

The breakup of Japan's NTT in July 1999 represented a turning point in telecommunications deregulation. The combination of remaining regulation and a lucrative customer base attracts arbitrage in the form of international callback services and Internet-based telephony services. As in other competitive markets, falling prices and corporate mergers characterize the domestic long-distance market. In 1997, Japan experienced a defining moment when the fixed-line telephony market declined by 11% at the same time there was an increase of 15% in the wireless market [15]. Wireless penetration was 31.7% in January 1999, driven by a near doubling of subscribers in their twenties. According to Japan's Ministry of Posts and Telecommunications, the number of mobile telephony customers exceeded fixed-line subscribers for the first time in March 2000.

Australian regulators have aggressively enforced the conditions to move the telecommunications market to full competition, by providing customer choice and giving competitors more direct access to customers. Growth in wireless has promoted competitive markets. Mobile penetration is high, as well as Internet usage. There are dozens of registered carriers and multiple mobile networks. Incumbent carrier Telstra has acted to position itself for the new marketplace. By the end of 1999, Telstra had achieved 40% cellular market penetration, assisted by healthy sales in prepaid mobile services. In high-speed data access, Telstra and Cable and Wireless Optus are competing fiercely for the high-income residential customer, creating duplicate hybrid fiber/coaxial cable facilities in some areas, while other areas remain unserved.

Liberalization initiatives are taking place in Singapore, Malaysia, and Indonesia, where the economic crisis has interfered with the development of its telecommunications sector. Singapore announced in 2000 that full liberalization would occur two years ahead of schedule and awarded 30 licenses to competitors of the incumbent SingTel. The country supports a dense fiber broadband network and a PC penetration rate of 40%. Nearly 20% of the country's population has two telephone lines at home. A partnership between SingTel Mobile, IBM, and MacroVision allows mobile telephone users to trade shares on-line.

Regulators introduced equal access in Malaysia in 1999, with five wireline operators in competition with Telekom Malaysia. Despite the economic climate, Telekom Malaysia and its competitors are investing in new wireline and wireless infrastructure. More than half of Indonesia's telecommunications service providers were facing possible bankruptcy resulting from the economic crisis in the region. Legislators took several initiatives to boost the sector, by relaxing rules concerning foreign investment and converting some operations between the state-owned PT Telkom and its partners into joint ventures.

According to CIT Publications, the number of cellular subscribers in Indonesia increased by 108% in 1999, to over 2.22 million customers. In 1998, before the Asian economic crisis was over, growth was only 16.4%. Market leader Telkomsel increased its market share to 46%, up from 40% in 1998. Service providers will likely launch 3G mobile telephony in 2004 or 2005, with interim standards and WAP applications available in late 2000.

Taiwan's high penetration rate of more than one telephone user per two people will help competition to blossom. Deregulation of most wireless technologies occurred in 1997, and fixed-line competition will ensue in mid-2001. State-owned LEC Chunghwa introduced a multimedia-on-demand service that provides movies, telecommuting, digital libraries, educational programs, games, and home shopping. In the first quarter of 2000, the Taiwanese government awarded three fixed-line licenses to compete with Chunghwa.

China is among the largest telecommunications markets in the world, and the country has made a commitment to upgrading its infrastructure, including the use of advanced foreign technology. China restructured certain aspects of telecommunications into multiple entities in the expectation that competition would foster development. While all service providers are state-owned, they compete against each other, creating a competitive market. Approximately 1.5 million Internet users were on-line in China in 1998, and the number of users is expected to rise to 6.7 million in 1999 and 33 million by 2003, for an annualized 60% increase [16]. China anticipates being the world's largest Internet telephony market, with more than one-third of its international traffic over the Internet by 2002.

2.5 Strategic positions

Incumbent access providers are positioned to succeed in a competitive market because of their strong brands, availability of capital, robust infrastructures, and existing customer bases. Their weaknesses, nonetheless, are important. First, while their access to capital is excellent, their financial structures can be a disadvantage compared to their newest competitors. The typical technology start-up is debt-free, capitalized well beyond the value of its assets, by investors who require neither short-term profitability nor dividends. The typical incumbent access provider has a high debt ratio—useful for a blue-chip stock, but not for an innovator in new markets—and supports an investor base that is risk-averse and dividend-hungry. High barriers to exit exist for low-margin markets. Regulators will not permit incumbent carriers to abandon unprofitable markets, and price increases in the interest of profitability are not an option.

New access providers face high financial barriers to entry (for facilities-based companies) or constant interactions with their suppliers (for resellers). The technologies they select are generally untested compared to the copper networks of incumbents. They do have the option of selecting the most promising markets for investment. The new carriers are not facing the disadvantages that confronted the first competitors in long-distance services at the outset of deregulation. Many of the newest challengers are well funded and are already giants in the telecommunications market, seeking to expand their business scope. Local providers will attack interexchange markets; long-distance leaders will move into local markets. Neither the technology nor the faces of their competitors are unknown to them.

Incumbent transport providers hold the advantage of decades of experience in competitive markets but lack the operations experience and local support structure of the wireline carriers. Companies will not succeed because of their previous market position. They will succeed instead through sound strategy and effective implementation.

References

[1] Masud, S., "Transforming the PSTN," *Telecommunications*, Vol. 33, No. 7, p. 22.

[2] Masud, S., "Mega Deals, Mega Risks," *Telecommunications*, Vol. 33, No. 3.

[3] Blau, J., "Europe's Window of Opportunity," *tele.com*, Vol. 4, No. 3, p. 47.

[4] Schmitt, J., "Churn: Can Carriers Cope?" *Telecommunications*, Vol. 33, No. 2, pp. 32–33.

[5] Blau, J., "To the Bone," *tele.com*, Vol. 4, No. 14.

[6] Harley, J., "Closing the Gap," *America's Network*, Vol. 103, No. 6, p. 26.

[7] Malim, G., "Competition Has Not Delivered Choice," *Telecommunications*, Vol. 33, No. 10, pp. 23–26.

[8] Beardsley, S., "Full Telecom Competition in Europe Is Years Away," *McKinsey Quarterly*, 1998, No. 2, pp. 32–37.

[9] Salz-Trautman, P., "Winning the Residential Race: The German Experience," *Telecommunications, International Edition,* Vol. 33, No. 11.

[10] Yankee Group Europe, "Competitive Operator Strategies: Different Strokes for Different Folks," Sept. 1999.

[11] Kuittinen, T., "Creating Mobile Models," *Telecommunications*, Vol. 33, No. 10, pp. 99–100.

[12] Masud, S., "Canadian Telecom: An Industry in Transition," *Telecommunications*, Vol. 32, No. 10.

[13] Morri, A., and J. Blazquez, "Neighborhood Watch," *tele.com*, Vol. 3, No. 7.

[14] Lerner, N. C., "Latin American Telecom Changes: Learning from Mexico and Brazil," *Telecommunications*, Vol. 33, No. 4, p. 25.

[15] Maamria, K., and N. Kobayashi, "Turning to a New Chapter," *Telecommunications, International Edition,* Vol. 33, No. 10.

[16] DeVeaux, P., "Internet Gone Global," *America's Network*, Vol. 103, No. 15, p. 32.

3

Competitive Successes

It is not the strongest of the species that survives, nor the most intelligent; it is the one that is most adaptable to change.

—Charles Darwin

It is surely much too soon to draw categorical conclusions about strategies in competitive markets. In the United States, the only unquestionably competitive telecommunications markets are wireless, interexchange service, and Internet access. For many local customers, especially consumers and those in rural areas, choices are less apparent. Some services, such as central office-based enhanced services, are simply unavailable to customers that do not support their own switch. Nevertheless, all telecommunications markets will change substantially due to advances in technology, such as the introduction of 3G technologies in wireless and a migration to a digital IP platform and the wider deployment of broadband services. Moreover, the interexchange market, as intensely competitive as it has been for a decade, will undergo a tremor when regulators lift the Bell company restrictions. European markets, because of the sequence of liberalization and other differences, are more competitive in the wireless segment and less competitive than the United States in local wireline markets. Pricing structures have migrated wireline usage to wireless communications, which is partly responsible for Europe's lead in introducing 3G technologies. Internet penetration is increasing as ISPs remove pricing barriers.

Most other telecommunications services markets are only in transition, in the United States and around the world. Some, like consumer local service, do not conform to competitive measures virtually anywhere. Therefore, judgments about successes, and for that matter, errors, are admittedly premature. In any case, telecommunications service providers have established strategies in the expectation that they would lead to growth. Some that have accomplished that objective in the short term are worthy of review. Other strategies have failed, due either to active misjudgment or simple negligence, and telecommunications service providers can benefit from the lessons that these failures teach.

3.1 Growth through mergers and acquisitions

Many telecommunications providers have actively sought out acquisitions to achieve growth. In general, acquisitions that are identified opportunistically to fill a product line gap, expand into a new geographical territory, or simply add capacity are not necessarily strategic. Acquisitions that increase the size of a company by a normal rate of growth, such as 10% or even 25%, are customary in most industries and can be viewed merely as an alternative to a significant buildout of facilities. A vast difference in size between acquirer and target removes many of the challenges of large mergers: allocating senior management slots, eliminating duplicate positions, or absorbing and managing cultural differences. Expansion through small acquisitions is simply an alternative to growth through normal business operations.

Other telecommunications service providers have used large mergers and acquisitions as the foundation of their growth strategies. Reports from A. T. Kearney and Mercer Management Consulting concluded that the majority of large mergers fails to create shareholder value [1]. Neither the strategy nor the price is a primary determinant of ultimate success, according to Mercer; the probability of creating value increases with superior postmerger execution. Furthermore, mergers between the largest companies, when the acquired company's revenues are at least 30% of those of the acquirer, have a failure rate of 75%. Telecommunications service providers are well aware of the risks of mergers and in many cases, have experienced failed mergers firsthand. Nonetheless, some telecommunications service providers continue to put their futures on the line when they choose to acquire or merge with companies that are leaders in their own right. In the telecommunications industry, several are notable for their ability to leverage their existing companies and grow at a geometric rate.

WorldCom

WorldCom was relatively unknown compared to its larger acquisition target when it set its sights on MCI in 1998. Surprising its leading competitors and most industry watchers, WorldCom managed to exploit its soaring market capitalization and climb the interexchange ladder from a distant fourth to a strong second place. The MCI acquisition was probably the most newsworthy manifestation of the WorldCom growth strategy, but acquisition had been the cornerstone of WorldCom's line of attack since its inception.

WorldCom began as long-distance provider LDDS and started its growth cycle through acquiring long-distance providers that were struggling in the intensely competitive market. Management's capacity to integrate and rejuvenate these companies produced an ability to buy additional candidates at a relatively low price. WorldCom's integration process would then create value and satisfy investors, sustaining its own market capitalization and beginning the cycle anew. Not all acquirers earn this market confidence. As WorldCom's market visibility increased, it acquired larger companies to expand its long-distance presence; its purchases included WilTel, IDB WorldCom, and MCI. Eventually, WorldCom needed to pay significant premiums as its targets became more sizable and appealing to other suitors. The purchase price for its proposed acquisition of Sprint included a substantial premium, but the strategic importance of the acquisition was invaluable.

WorldCom used acquisitions to expand its scope as well as its scale. Recognizing the need for a data network, WorldCom acquired leading provider UUNet and CompuServe Network Services. To gain facilities-based local access, WorldCom acquired MFS, among the top CLECs, and acquired additional facilities from Brooks Fiber. Its bid for Sprint proposed to position WorldCom among the top telecommunications service providers in the long-distance, local, and wireless markets. Until new entrants (such as RBOCs) enter the market, only AT&T is able to make that claim in the United States.

The stock market and WorldCom's customer base continue to receive the company's acquisitions well. While it is too soon to claim categorical success in the competitive arena, most observers would agree that finding and completing acquisitions is a core competence held by WorldCom, in a market that values large size and leveraged resources.

WorldCom's strength is in its ability to focus on markets it chooses to serve and to ignore opportunities in markets that are out of its scope. The company's first local-service-provider acquisition was MFS, a company that built fiber optic rings around business communities. At the time, WorldCom was avoiding building or buying facilities to serve consumer markets, which did not fit with the company's strategy to select markets that will become profitable quickly. Consumer markets are large and wide-ranging, and it is nearly impossible to target specific segments efficiently. Carriers that choose to serve consumers must be willing to serve unprofitable customers.

WorldCom's choice of acquisition as a growth strategy recognizes its high cost relative to buildout, but the company has decided that rapid entry into desirable new markets is worth the premium. Its record for integrating acquisitions has bolstered its stock price, which alleviates the premium most companies need to present when buying into new markets.

SBC

SBC's postdivestiture statements that it planned to "stick to its knitting" prompted observers to assume that the company would remain in its traditional businesses and within its regional footprint. The misinterpretation was understandable, as most of the SBC area is rural, and the company's management, until divestiture, had been even more conservative than the rest of its RBOC siblings. SBC did not make the first acquisition among the divested RBOCs. Nonetheless, the company gained prominence and respect when it made the first major acquisition in the industry—and the first significant acquisition out of region—with its purchase of Metro-Media, a nationwide cellular provider. Until then, three years after divestiture, local telecommunications providers had not made inroads into out-of-region services. In 1996, SBC's acquisition of Pacific Telesis represented the first merger of two divested RBOCs. Later, SBC continued its local service expansion, with acquisitions of Southern New England Telephone (once a small AT&T subsidiary serving Connecticut) and Ameritech, another divested RBOC. With all the acquisitions completed, SBC serves about one-third of the access lines in the United States, and with equally aggressive acquisitions in the wireless market, SBC was in third place in the United States behind AT&T and Vodafone AirTouch. The company has also made significant investments in service providers outside the United States.

The key to SBC's success is that it has indeed "stuck to its knitting." Its acquisitions have focused on businesses in which it has experience in both the technology and the management. Like WorldCom, SBC's approach to integrating acquisitions is not a particularly democratic one. Still, completing the Ameritech integration carries additional conditions imposed by the FCC and by SBC's own strategic plans. First, the FCC's requirement that SBC enter 30 local markets outside its own territory and offer service competitively will introduce significant market risk and unprecedented management requirements. Separately, SBC made a commitment to invest heavily in DSL technology to offer to its local customers. Both of these initiatives will demand both capital and management attention. If it succeeds in both implementations, SBC will prove that it can meet its goals of staying within its established businesses while becoming a giant in the industry.

Bell Atlantic

Bell Atlantic announced the first acquisition of any RBOC within days of divestiture in 1984, and its pace and scale have accelerated over time. Bell Atlantic has achieved the same eventual position as SBC, but it took a different route, both strategically and operationally. Bell Atlantic's apparent initial strategy was to integrate horizontally through local cable services. Its most dramatic announcement of a merger involved a coupling that eventually dissolved with TCI, a local cable provider ultimately acquired by AT&T through its own local cable strategy. Bell Atlantic's later mergers have been as substantial but not as sensational as its TCI announcement. Accordingly, the company, possibly unnerved by the challenges of integrating a local service provider with different technologies, different operational processes, different regulatory influences, and perhaps a different corporate culture, changed its acquisition strategy. Its most recent mergers have increased its local footprint in much the same manner as SBC. Bell Atlantic's merger with NYNEX and its combination with GTE as Verizon have made the company the largest local provider in the United States, as measured by access lines.

Operationally, Bell Atlantic's approach to mergers has been slightly more deliberate than those of WorldCom or SBC. The company accomplished its NYNEX merger in a series of small steps. First, Bell Atlantic and

NYNEX established a marketing alliance between their respective wireless divisions. Because wireless is by nature a transient service, the larger the coverage area served by a provider, the more flexible the terms of service and the more valuable the service is to traveling customers. Customers of the Bell Atlantic/NYNEX alliance could enjoy seamless service when moving between the two territories. Bell Atlantic and NYNEX also achieved an advantage: the ability to integrate their marketing and operations. This arrangement enabled the companies to leverage their marketing resources and reduce their costs. Strategically, it offered the ability to merge the wireless companies into a jointly owned entity. The wireless venture operated under the management of the two companies, and soon afterward, Bell Atlantic and NYNEX merged their entire operations. Bell Atlantic's partnership with Vodafone AirTouch presages a more serious commitment in the future. First, a marketing alliance strongly resembles the NYNEX agreement in its earliest stages, and second, the company markets the large wireless footprint using the same name, Verizon, as the merged GTE/Bell Atlantic entity.

This approach undoubtedly requires more negotiation because it occurs over a longer period, but using a marketing alliance as a precursor to a merger (and in this case, merging divisions before merging completely) offers several advantages over a sudden merger agreement. The obvious advantage is the ability for both companies to learn about each other and adapt outside of the scrutiny of official merger integration. Companies can begin to achieve economies and incorporate the best operational practices within each unit into an overall design. Externally, the alliance somewhat serves as a barrier against takeovers of either partner by outsiders. The barrier is permeable, though, as British Telecom learned when it lost MCI to a WorldCom offer despite a similar arrangement.

Bell Atlantic was bolder in its offer for GTE, with which it had no significant alliances when it agreed to merge in 1998. The merger was labeled "a merger of equals"; indeed, a GTE executive leads the management team, and the company name has changed to reflect the nationwide (or global) scope of the resulting company. The GTE merger, unlike a union with another RBOC, represented a change in business scope as well as scale. Unlike the RBOCs, GTE is unrestricted from offering long-distance service to its own local customers. This does not pose a constraint on the merger, as Bell Atlantic is free to offer long distance outside its divestiture-assigned territory. The acquisition positions Bell Atlantic for the eventual lifting of the restraint, with a national network, a marketing organization, and other infrastructure to support an interexchange business. GTE's ownership of data network BBN Technologies, which owns a nationwide Internet backbone, would have positioned Bell Atlantic in another strategically important market, but GTE's ownership of the unit was greatly reduced as a condition of merger approval. Still, GTE has extensive experience in operating data networks, and much of that knowledge will stay with the parent company until it is permitted to return to the backbone business. Therefore, the GTE acquisition offers Bell Atlantic expansion in both size and scope.

Bell Atlantic's acquisition strategy has generally directed it to lower-risk acquisitions, in its business reach and in its willingness to pay an acquisition premium. Neither the NYNEX nor the GTE merger involved significant premiums; in both cases, the merger partners were gaining benefits commensurate to those obtained by Bell Atlantic. Premiums increase the already substantial hurdles between the negotiation of purchases and their successful integration. Investing in the search for willing acquisition partners is a Bell Atlantic core competence, making it more likely that acquisitions will be worth their cost.

3.2 Pricing

The primary pricing strategy currently favored by telecommunications service providers is to reduce prices as much as possible. A secondary strategy has been to manage network utilization by offering incentives to place calls outside of active business hours. These techniques are effective, but unoriginal. Furthermore, competitors can easily reproduce these initiatives, so they only maintain or reduce every service provider's revenues with a small increase in volume. Innovative pricing has rewarded service providers with market share and higher revenues.

MCI Friends and Family

MCI enjoyed a significant lift in its market share with its introduction of the Friends and Family campaign. This program demonstrated the company's ability to innovate using both pricing strategies and technology to develop its market. Furthermore, the campaign is among the most successful and dramatic demonstrations of success through excellent strategy in the telecommunications industry.

MCI began its long-distance service in the business market, by offering discounted services between points in Chicago and St. Louis served by its microwave facilities. Many of the company's earliest challenges were in the regulatory arena. Nevertheless, by the time MCI moved into the consumer market, it needed to find ways to build its share against AT&T, an incumbent provider with a staunchly loyal base. The interexchange market was by this time becoming quite competitive, with hundreds of resellers offering discounted service. Long-distance service was (and still is) largely a commodity to consumers, who are highly receptive to low prices. Offering low prices without additional incentives for customers, however, led to three long-term problems. The first is that price wars decrease profits for the entire industry. The second is that service quality is important to consumers, who have few opportunities in advance to test services to decide whether the quality of a low-priced service is adequate. Excessively low prices tend to scare consumers away rather than attract them. Resellers tend to have very low-cost distribution networks and little advertising expense, so a national branded carrier like MCI could not always match the price of the lowest-cost competitor. Third, customers acquired through promotional pricing are the likeliest to churn when a competitor offers a more enticing price. In an industry of declining costs, competitors dangling lower rates are ubiquitous. MCI needed to find a point of differentiation beyond lower price.

Its solution was the Friends and Family service, in which consumers received a discount on calls between themselves and other MCI customers that they identified in their "calling circles." This had several benefits to MCI: It provided an attractive discount to customers without positioning the brand as a cheap or lower quality long-distance alternative. It also encouraged customers to act as sales representatives and recruit new customers to include in their own calling circles. Last, in an industry characterized by high churn, the calling circle discount served as a significant switching cost. If one customer left MCI, its chief competitor AT&T would not be willing to match the Friends and Family discount. Moreover, the customer's change of carriers affected not only the price the customer paid but also the price that their friends and family would need to pay to reach them. As an added benefit to MCI, AT&T could not respond competitively to the Friends and Family program, either strategically or tactically.

Strategically, AT&T was unable to create a similar program. With market share of about 80–90% of consumers at the time, AT&T would need to include virtually every consumer on someone's list. This would create a virtual across-the-board discount to consumers, as most customers already used AT&T as their primary carrier. MCI customers were unlikely to recruit everyone they knew to MCI. Each resulting personal calling circle included a small percentage of the market and accordingly included only a small percent of MCI calls, no matter how successful the program would be. Calls to customers outside the calling circle (that is, nearly everyone) were charged at the full MCI long-distance rate.

Tactically, AT&T could not respond in any case. Tracking the customer base and ascertaining which calls qualified for discounts were well beyond AT&T's legacy call measurement systems. The sheer volume of AT&T customers that qualified for calling circles would have created a huge tracking operation, a problem not shared by MCI and its small database of customers. Furthermore, as a new operator, MCI had access to the latest information technologies, including relational databases. MCI could design its billing systems with much less effort and fewer lingering requirements than it would take AT&T. In short, MCI used its weakness—being a small competitor—as a strength that increased its market share dramatically. It attacked AT&T on its own strength, its large market share, and on its weakness, the outdated information technologies that supported its billing operation.

Later, MCI (now part of WorldCom) de-emphasized its Friends and Family program in favor of more conventional discounts. As its market share grew and new share growth was slower, the company focused its energies on customer retention as well as customer acquisition.

AT&T Digital One Rate

AT&T, as the market leader in long-distance services, is not expected to initiate innovative pricing solutions. Its strong brand in its markets offers some price premium over equivalent services offered by less venerable competitors. Market leaders often retain share as price followers. Indeed, AT&T rarely innovates in long-distance pricing. AT&T's long-distance pricing strategy most often follows pricing initiatives announced by Sprint and MCI WorldCom. As the market leader supporting a customer base characterized by inertia, price leadership is not necessary and can be risky. On the other hand, in the wireless market, AT&T was a late entrant. Wisely, the company revised its pricing strategy to meet the needs of its market.

AT&T seized its strategic advantage when it created the Digital One Rate program. The program eliminated most of the numerous charges paid by the targeted high-value customer. Customers were used to paying for local airtime in their home territory and a combination of long-distance charges, flat daily roaming charges, and high airtime charges while roaming outside the service provider's local territory. With the array of costs, many unpredictable and without protection from charges from incoming calls, many traveling mobile customers were reluctant to use their mobile telephones or even turn them on while outside of their home territories. This resulted in the ironic behavior of customers using the telephones less when they were mobile than when they were near their home landline. AT&T's Digital One Rate program eliminated all of these charges, including long distance, an unprecedented action in the United States, and replaced them with a single package of minutes, such as 600, 1,000, or 1,400 minutes per month. The packages were each priced so competitively that many per-minute charges were lower than the price many AT&T customers paid for a long-distance call on a landline. At the time AT&T launched Digital One Rate, the 1,400-minute package came to 11 cents a minute, for customers whose AT&T discount program was 15 cents per minute on their landline. Some customers opted for the large package of minutes and transferred their long-distance calling to the wireless network, if only to consume the minimum usage and ensure steeper discounts.

AT&T's objective was to capture the target customer segment and to move customers to its digital network. To prepare for the service offering and its own goal of a nationwide footprint, it obtained PCS licenses in important markets and set up partnerships with wireless carriers in markets it considered strategic but did not serve. AT&T developed a national roaming database and upgraded its network, worked with the handset manufacturer to manage the specific program requirements, and developed an advertising campaign covering the areas where it planned to offer the service.

The service generated 200,000 new customers in 60 days, most of which were lured from other service providers [2]. Moreover, AT&T led a 1999 J. D. Power survey for customer satisfaction in nine of the 13 markets surveyed. In the short term, its competitors, most of whom were regional providers without national coverage or a nationwide long-distance network, were unable to compete with the service offering or the pricing. Even if the competitors could have put contracts with other regional carriers in place, many wireless providers are RBOCs that did not support national facilities-based interexchange networks. Revenue-sharing arrangements would also create costs above competitive prices. With

AT&T's nationwide footprint, the Digital One Rate program was simply unassailable.

In the long term, the program changed the structure of pricing. AT&T's pricing strategy, even for long-distance services, is to keep the price simple. Ironically, AT&T is the company that originally put the complex monopoly pricing systems in place, such as rating calls based on actual distance or time of day. In the competitive world, AT&T has found success in straightforward pricing plans. In the wireless market, the first competitive change that AT&T caused was to simplify pricing for all wireless providers, to the degree that they could afford to do so. The second, more permanent change was to hasten the consolidation in the industry. Surviving companies with regional networks immediately sought out merger candidates or marketing partners. Large partnerships and mergers, such as the merger between GSM providers VoiceStream and Omnipoint, were undoubtedly influenced by the need to develop a nationwide footprint as soon as practicable. Bell Atlantic's failed gesture toward acquiring AirTouch reincarnated itself into a partnership with AirTouch's new parent, Vodafone, with the stated objective of achieving nationwide coverage.

Sprint

Sprint has inhabited third place among U.S. IXCs for most of the industry's competitive history. Sprint has acted as the pricing pioneer for the interexchange industry. In 1995, Sprint introduced per-minute, distance-insensitive long-distance rates, reducing rates, and more importantly, simplifying the purchase decision for customers. In 1999, Sprint introduced a flat per-month, unlimited weekend calling rate, the first mainstream provider to present a true flat-rate service. Validation of the strategic decision occurred when AT&T offered to match Sprint's rate.

In all cases, Sprint's pricing innovations have been original, but cautious. The unlimited weekend package transfers consumer minutes of use from times of normal traffic to times of network underutilization. Sprint benefits from the stabilization of traffic and reduces its call measurement and billing costs; consumers benefit from lower prices and price simplification.

Sprint has innovated in business pricing as well, to attract the cost-conscious small business customer. Its Fridays Free campaign encourages businesses to use its service with the incentive of free long distance one day per week. Sprint recognized that most businesses cannot delay making important calls to save a few cents and that the incentive of free calling would not shift an unreasonable amount of traffic to Fridays. The promotion simply enhanced Sprint's market share among the cost-conscious small business segment. Another business offering, Sprint Business Flex, enables customers to configure their own integrated low-cost service bundle.

In its role as the third-place contender, Sprint is expected to offer low price as an incentive to win customers. Nevertheless, with its distant third status, and the intense competition between facilities-based carriers and resellers, Sprint cannot risk becoming a low-cost alternative with low price as its only point of differentiation. To ensure that its pricing draws customers, Sprint underscores its market position among industry leaders with two other efforts. First, Sprint maintains technology leadership with significant investment in its fiber network. Second, Sprint educates customers about its brand and about the criteria that customers need to use to select a long-distance provider. Not surprisingly, the attributes it suggests customers use to select a long-distance provider are those at which it excels.

To stake a claim to technology leadership, Sprint built the first all-digital fiber optic network in the United States. The company routinely invests in developing its network and maintaining leadership in its technology platform for voice and data services. Sprint will undoubtedly be among the first, if not the first, long-distance provider to announce and operate a nationwide local and long-distance IP network.

Sprint makes large investments in advertising to educate and persuade potential customers. The size of its advertising budget approaches the industry leaders in its absolute size. In 1998, Sprint's advertising of $462.4 million made it the ninth most advertised brand of any kind in the United States [3], coming close to the $550.8 million spent by AT&T and the $636.2 million MCI WorldCom advertising expenditure. Against their respective market shares, Sprint spends nearly twice MCI WorldCom's expenditure for each point of market share and almost four times that of AT&T. Sprint has reaped the rewards of successful branding for its investment. Its "pin drop" campaign generated valuable recognition for the brand and the network quality that Sprint wants to convey. For four consecutive years, customers named Sprint as the top long-distance provider for high-volume users in the J. D. Power and Associates customer satisfaction studies. The company's constant advertising reminders of the high-quality network were almost certainly a factor in customer perceptions.

3.3 Embracing competition

Some countries have demonstrated that the smoothest, or at least the fastest, transition to competition occurs when regulators and service providers act as partners. It is possible already to see widening gaps in prices, large differences in the penetration of new technologies, and significant cultural change within the incumbent provider's management among the world's largest providers. In countries where incumbents fight deregulation, prices remain high and choices are limited. This is the most defensive strategy but the least likely to gain long-term success when competitors break through the market borders. Others looked farther ahead and developed competitive strategies that included risk taking and innovation. The incumbent service providers described in this section succeeded by embracing competition rather than contesting its minutia at every opportunity.

Finland

In 1994, Finland opened local and long distance to competition; it deregulated international communications in 1997. The incumbent carriers were Telecom Finland (now called Sonecor) and a coalition of local carriers called Finnet Group, whose members covered 75% of the landmass and 25% of the population. Telecom Finland held the monopoly on long-distance service. The presence of two relatively evenly matched incumbent competitors created an immediate burst of competition. By the end of the first month of competition, Telecom Finland lost half of its long-distance market, which had accounted for about 50% of its revenues. Within two years of competition, the company had lost 60% of its share.

Deregulation was early and swift. The presence of two strong providers had an influence on the rapidity of market entry, but the service providers created the competitive marketplace through their own initiatives. Finland's service providers did not choose, as so many American incumbent providers do, to salvage their waning market shares in the courtroom or through leisurely planning and deployment. Instead, they chose to embrace the new arena and win. Prices for long-distance calling dropped by 75% in three years. According to Invest in Finland, telephony and data services in the country are among the lowest priced in the world. Competitive intensity in the business market encouraged service providers to compete in the consumer market sooner than has been the experience of other jurisdictions. For the first time anywhere, Finland consumes more minutes on its wireless network than its wireline one [4]. Finland's early adoption of data technologies created an understanding of data services, which has resulted in high Internet penetration.

Finland can boast many firsts in European communications development. Finland established Europe's first electronic mail system in 1986, built the first GSM mobile network in 1991, instituted the voice-over-Internet protocol (VoIP) in 1996, and became the first country where four of every 10 people had mobile phones [5]. More recently, Finland has reported mobile penetration rates of 60% in some areas [6]. One of Finnet's members offers a service in which customers can use a wireless telephone to receive calls anywhere in Helsinki, for the wireline rate, and make calls at a discount from the normal mobile rate.

The combination of strong competitors with progressive regulators, a technologically capable population, and swift implementation has made Finland a model of deregulation. Strategies used by incumbents include customer-focused deployment of new technologies, aggressive price and cost cutting without creating long-term price wars, and management's willingness to take risks.

Chile

In the early 1980s, Chile was among the first telecommunications markets to undergo deregulation, and the country now serves as the Latin American leader in competitive services such as long distance and wireless. In local service, Compañia de Telecomunicaciones de Chile S.A. (CTC) still controls 90% of local lines, an improvement over most world markets, but not yet competitive.

Chile's role as trendsetter is especially noteworthy in light of growth in telephone penetration from 6.5 lines per 100 inhabitants in 1990 to about 18 lines seven years later. The world average penetration is about 11.5, making Chile a stronger than average country. Other than Singapore and New Zealand, Chile is the only country in the world with an all-digital network, completed in 1993. Subscriber growth in mobile markets was 126% during 1998, despite an economic downturn [7]. According to the Organization for Economic Cooperation and Development (OECD), telecommunications competition resulted in a reduction of 36% in rates for local calls between 1989 and 1994, 38% for long-distance calls, and 50% for international calls. Competition has sparked a technology race among the top four service providers, which have each installed fiber optic transmission networks, sophisticated switching platforms, and fiber rings around Santiago [8].

Chile liberalized long distance in 1994, and prices for domestic and international services dropped dramatically. The ensuing shakeout was more dramatic than most markets would choose to undergo, and all but three providers disappeared under the intense price competition. At the height of competitive intensity, calls to the United States were as low as \$0.02–\$0.05 per minute. The price eventually returned to a more sustainable yet competitive rate. Meanwhile, the relatively high cost of Internet service has contained expansion in data services. Growth has increased in the last five years because of a proliferation of competitors, an increase in Spanish-language Web pages, and the consolidation of the industry.

Chile owes much of its success to the willingness of regulators and competitors to work together and innovate in anticipation of stimulating usage. For example, CPP, implemented in 1999, should create new usage in the mobile market as it has in other markets that have adopted the practice. In addition, long-distance carriers are working together to develop a unified customer bill. Unlike their counterparts in the United States, customers in Chile select their long-distance carrier on a call-by-call basis by dialing a three-digit code. Customers can use any of 12 interexchange networks and receive billing at the end of the month. A great deal of coordination among carriers is required, including the investigation and resolution of uncollectibles. Liberalization in Chile demonstrates that competition can thrive in less-than-ideal economies, without pent-up demand for telephone penetration, and without significant regulatory incentives. In Chile, competitors simply wanted to compete.

References

[1] Masud, S., "Mega Deals, Mega Risks," *Telecommunications*, Vol. 33, No. 3.

[2] Albright, P., "Stunning Pricing Shakes It All Up" (*Wireless Week* 1998 Carrier Excellence Award).

[3] "Top 100 Megabrands 1998," *Advertising Age*, Vol. 70, Iss. 29, p. S2.

[4] Bartlett, T., "Do Mobile Services Provide Consumer Value?" *Telecommunications*, Vol. 33, No. 1, p. 27.

[5] Evagora, A., "The Nordic Track," *tele.com*, Vol. 3, No. 1, pp. 68–72, 75.

[6] Beardsley, S., and A. Evans, "Breaking the Access Bottleneck," *Telecommunications, International Edition,* Vol. 33, No. 7.

[7] Neal, C., "Setting the Latin Pace," *Telecommunications, International Edition,* Vol. 33, No. 10.

[8] Connolly, C. F., and H. Hernandez, "Emerging Economies," *tele.com*, Vol. 2, No. 11.

4

Blunders in the Competitive Market

We cannot sell our listings; they are too valuable.

—RBOC "Yellow Pages" marketer

Telephone books on CD-ROM first arrived when NYNEX, later part of Bell Atlantic, marketed a directory of the New York area. The company charged $10,000 per disk and sold copies to government bureaus. The executive in charge of the product left to set up his own company but could not get his former employer or the other local service providers to rent the computerized listings for a reasonable price. Instead, the new company hired workers in Beijing to type listings into a computer (twice for error correction) for $3.50 per day per worker. The resulting product (for the entire country, not just New York) cost well under one dollar apiece to produce and sold profitably for hundreds, not thousands [1]. The above quote about listings being too valuable to sell, incidentally, is from a product manager at a different RBOC, so NYNEX's philosophy was not an anomaly. An ingrained monopolist mentality undoubtedly assumed that customers who could not buy what they desired from the monopoly provider would simply do without it.

Every successful telecommunications provider needs to take strategic risks, and thus blunders are inevitable. Those that take no risks are making a strategic error of omission, inviting failure from the opposite direction. Some technologies fail once or several times before they succeed. Interactive television is probably an example of this.

Strategic errors and misjudgments are highly visible, but ideally, their duration can be as short as management's alertness to recognize them and willingness to resolve them. Many of the strategic missteps identified in this chapter were resolved and their consequences minimized. Others represent the core values and business processes inherent in company cultures. They will continue to surface in assorted forms until management approaches change fundamentally.

4.1 Positioning and pricing

Some service initiatives failed because the service provider did not define its target market, consider installation and the service itself from the customer's perspective, or recognize that competitive markets operate differently from monopolies. Competitors, unlike monopolists, do not control service introductions, prices, or the availability of competing products. Service providers that fail to consider the competitive impact of their offerings will not meet their marketing targets.

Integrated services digital network

ISDN offers high-speed data services over the local copper loop. Dubiously supplanting AT&T's experimental PicturePhone of decades ago, ISDN has served as an illustration for failures of technology and marketing in the United States. Apologists for its disappointing service penetration argue that ISDN was simply ahead of its time. Nonetheless, other technologies have managed to create their own markets and drive demand in the United States, and ISDN has been successful outside the United States. Additionally, the rationale of customers not being ready does not explain the present unmet demand for broadband access, nor does it account for the multitude of customer complaints about the ISDN experience from early adopters. Obviously, willing customers were present.

In Europe, ISDN has been more successful, even in countries where interest in Internet connection occurred later than in the United States.

Figure 4.1 demonstrates that the United States, with Internet users comprising 37.4% of its population [2], leads most of Europe in Internet penetration. The dearth of ISDN users in the United States is not attributable to lack of a market.

ISDN represents about 10% of Telecom Italia's customers [3]. British Telecom, perhaps in response to the introduction of cable modem service, launched an aggressive campaign for consumer ISDN. Germany's Deutsche Telekom claims the world's largest ISDN network. This is as much a testament to the service provider's marketing focus as anything else. Germany's Internet penetration is average, so demand is not driving ISDN growth. Only in the last few years, when there has been more broadband demand than ever and when superior or less expensive alternatives have become available, has ISDN become even an adequately viable market in the United States. Nevertheless, ISDN will not survive when other broadband technologies are widely deployed.

ISDN missed its market potential for a host of reasons, including the following:

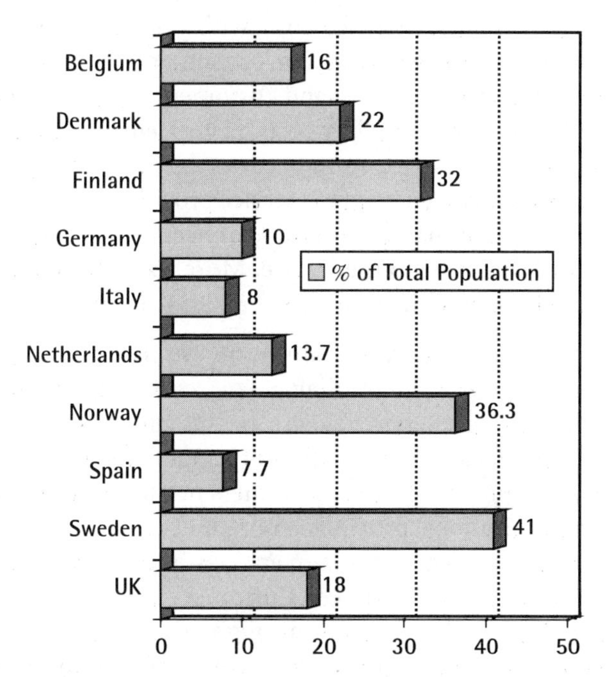

Figure 4.1 Europe's use of the Internet.

- *Customers did not understand the technology, and service providers failed to educate them and create demand.* Unlike other advanced telecommunications services directed at consumers, telecommunications services providers could not demonstrate ISDN on a television commercial. Even the name failed to sell the service's benefits; it simply described its architecture, and the phrase "integrated services digital network" contains three adjacent and relatively meaningless modifiers. BellSouth is now selling ISDN as "incredibly speedy data now" in an effort to market the service's benefits instead of its technology. For business customers, service providers might have feared that an explosion in ISDN sales could erode the profitable T-1 market. This would not be the first time that monopolists missed a market opportunity for fear of cannibalizing existing services. This is an especially problematic cultural hurdle in competitive markets, where service providers do not control the array of alternatives available to customers.
- *Installation was expensive and disruptive.* ISDN required one or several trips to the customer location by a technician. In keeping with pricing policies that matched costs with prices, customers paid for each service technician visit. For consumers, telecommunications service provider visits can strand customers in the house or require time off from work. Ancillary costs, historically uncommon to telecommunications services, annoy customers, and charges outside of those already incurred from the service provider, such as modems or increased on-line charges for faster connections, added to the disincentives to subscribe to the service. More recently, developers have eliminated much of the initial costs and inconvenience.
- *Some service provider technicians were not as knowledgeable about ISDN as they were about traditional services.* Perhaps because providers deployed the service slowly or because the market lacked momentum, local providers did not support the technical installation and support teams that customers needed and expected. Local telecommunications providers, not outside teams, installed and managed the service. There is an analogy to the two ways customers can upgrade to DSL service. Customers can obtain DSL either through their local provider or through a data CLEC. The advantage of buying DSL from a local service provider is that the customer has

a single supplier. The advantage of using a data CLEC is that the technician is a specialist in the technology.

- *Pricing for the service reflected needs of the service provider rather than customer demand.* Telecommunications service providers used historical guidelines to price ISDN service. Rates matched costs in terms of level and timing. It took years for providers to realize that one-time charges and usage fees discouraged customers from subscribing to the service. Eventually, service providers waived up-front charges and sought simple installation procedures that no longer required as much on-site technical support. Rates for ISDN finally fell because of competitive pressures from comparable offerings such as DSL and more likely because of immediately available competitive alternatives such as cable modems. Without the prospect of customer loss, it is difficult to predict when the problems of ISDN would have been resolved.

Two issues should still concern the management of incumbent local service providers. The first is that most of the reasons for ISDN's disappointing performance were due to monopoly management of a competitive service. Local providers controlled the timing of the deployment, a luxury available to monopolists, but not competitors. Pricing for ISDN reflected historical accounting methods, rather than pricing the service based on its value to customers. Management often extolled the service as a source of growth and profit but took no risks and made few investments to ensure that the service would succeed. Confronting imminent competition in every arena, management needs to eliminate monopolistic approaches to service deployment. Competitive markets are unforgiving.

The second issue of concern is that DSL deployment is following an improved but similar path to that of ISDN. Telecommunications service providers have learned some marketing lessons about service names and pricing, but deployment is slower than it needs to be. Telecommunications service providers sometimes treat data CLECs and their marketing partners as competitors rather than as customers or service partners in the provision of DSL to customers. Channel conflict is partly responsible for this dilemma. Telecommunications service providers that learn from failures such as ISDN will develop strong business models and core competencies to overcome these obstacles.

Iridium

Iridium began in the late 1980s as a joint venture with deep-pocketed investors including Motorola. Its original objective was to build a network of 66 satellites that would provide wireless telecommunications service anywhere on the globe. The service targeted business customers in remote regions, where wireline connections were scarce and service would be valuable. The target customer needed to be willing to invest thousands of dollars in the telephone itself and pay for access at a multiple of international calling rates. When the company filed for bankruptcy, the venture had invested $5 billion, yet it had delivered only 10,000 to 20,000 of the 50,000 or more customers predicted. At the end of its service in March 2000, it ended service for 55,000 customers.

The failure of Iridium is attributable to several causes, including the following:

- *The price was too high to find a large enough market.* The initial fixed price for the telephone was $3,000; it later decreased to $1,500. A large unit price often deters potential customers from subscribing in the first place. With a new technology and unknown applications, customers are often reluctant to invest. For example, wireless providers initially amortized the high price of the cellular telephone in the monthly contract rate for the first year or two of service. The Iridium business plan apparently did not consider the rapid increase in conventional wireless technology coverage, nor did it anticipate the significant price decreases for wireless as PCS services created new competitors for traditional cellular providers. Traditional wireless service is not a substitute for satellite communications, but it serves as a benchmark. These trends in standard wireless service created a larger price gap between Iridium and its closest competitors, especially considering the elimination of roaming and long-distance charges, which would benefit Iridium's target customers disproportionately. Customers who would make a satellite-based call, rather than wait for service within their home wireless network, might reconsider when a large price disparity is present. Iridium later cut its airtime prices to reduce the gap between its service and its nearest competitors.
- *Time to market is essential when competing technologies are present.* The aggressive schedule for satellite launches did not adequately

account for technical problems and weather delays. Missing business plan deadlines worries investors, but it also concerns potential customers, who are hesitant to make the investment in a service they fear will not last or will not operate where they need to use it. Moreover, in spite of its high costs, the service had normal technical difficulties after it launched. This too concerned customers and investors. Furthermore, the capabilities of the service lagged the features of ubiquitous wireless service. For example, Iridium did not offer data services, and it did not work well indoors. Data services are becoming critical as the Internet becomes a standard communication and work tool. For some worldwide communication, access to e-mail, because of its time-shifting ability, is more important than voice communications.

- *Not enough telephones were available when the company launched the service.* This is a more serious error in planning than it appears. If customers had subscribed to the service in the numbers expressed in the business plan, or if demand exceeded expectations, the company would have endured customer backlash from its inability to meet demand. Some customers complained that they found the telephones difficult to use. High-price services require extraordinary customer service. In fact, Iridium created customer care to meet the unique needs of its customers, supporting several virtual call centers, including access number-directed language support in 13 languages, and access to 40 others [4]. Still, part of customer service excellence is the ability to anticipate change in the forecast.

- *It failed to exploit competitive advantages in niche markets.* Government represents an excellent target market for an independent worldwide wireless network. The service does not rely on switched local telephone systems, so it possesses military advantages over its wireless competitors. Journalists, aid workers, and refugees in the Balkans used donated telephones during the 1999 military action. During the product launch, this vertical market would have bolstered the numbers of subscribers and produced revenues to the venture. Other vertical markets might have been more successful with a more targeted application of marketing and technical efforts. Iridium's later activities had more focus than the original project.

AT&T's channel conflict

Long before AT&T and the U.S. Department of Justice resolved the long-standing antitrust suit against AT&T, one solution proposed was for AT&T to divest its manufacturing arm, Western Electric. AT&T resisted that alternative, and when it divested its operating companies instead in 1984, Western Electric remained with the parent company. AT&T undoubtedly believed that vertical integration, the ability to manufacture its own network switching equipment, was critical. Indeed, several of North America's telecommunications service providers, including Bell Canada's Northern Telecom (Nortel) and GTE's Automatic Electric, made the same strategic investment.

After divestiture, and as AT&T's long-distance market share slipped from 90% to 50% of the market, the company realized that lost equipment sales to other IXCs were outpacing the strategic advantage of vertical integration. AT&T began to experience the consequences of channel conflict.

Channel conflict results when a company's distribution channels intersect with those of its customers. One example of channel conflict is widespread due to the growth of the Internet. Many companies support e-commerce on their own Web sites but maintain relationships with dealers and agents. The result is that a company, such as a manufacturer, sells products directly to end users in competition with its core customers, the agents and dealers. Not surprisingly, those downstream in the distribution channel resent the competition from their own suppliers. Solutions are achievable, but they require active cooperation between supplier and distributor.

In AT&T's case, competitive IXCs were reluctant to purchase AT&T network switches, knowing that profits from switching equipment would support AT&T's competitive activities in long-distance services. Local Bell companies, equipment customers of AT&T through tradition and inertia, feared AT&T's imminent status as a competitor, once the company offered local service and the regulators lifted RBOC restrictions against long-distance service. Customers also feared that proprietary information and plans that they provided to their equipment supplier could become available to their competitor in the telecommunications services market. An indirect consequence of channel conflict is the distributor's uneasiness that a supplier that is also a competitor wants the distributor to fail, or at least is not totally committed to a distributor's success. In a competitive market, this is a significant detriment.

AT&T resolved the channel conflict when it divested its manufacturing division, now called Lucent Technologies. AT&T apparently decided that the opportunity costs of outside markets were higher than the cost of lacking an in-house manufacturer. Both surviving companies are thriving, and the channel conflict is gone. There are lessons in AT&T's experience for other telecommunications service providers, including AT&T. In both long-distance and local markets, companies are acting simultaneously as wholesale and retail channels. Facilities-based IXCs provide minutes to resellers. If these carriers consider their wholesale operations to be an important strategic element of their businesses, channel conflict is a problem for them. Wholesale providers such as Williams Communications offer communications capacity to resellers without also presenting their customers with competition.

AT&T's experience can be instructive to telecommunications providers in the interexchange and, especially, local markets. Several local providers began making overtures to separate their wholesale and retail businesses. In the late 1990s, both Frontier Corporation and Southern New England Telephone (SNET) restructured their operations into wholesale and retail operations. Later, Global Crossing purchased Frontier, and SBC acquired SNET. Neither parent company has declared a strategy for wholesale or retail operations. Frontier, in fact, is Global Crossing's first acquisition with a significant retail base. The company's attempt to acquire US West at the same time (losing the skirmish to Qwest) implies that Global Crossing is now pursuing a retail strategy. Whether it will focus on retail or combine its retail holdings with a wholesale business model is not clear. British Telecom announced a similar restructuring.

RBOCs continue to offer both retail and wholesale services, often in separate organizational divisions. Regulators require their wholesale activity, whether profitable or not, but most RBOCs insist that their wholesale businesses are robust and strategically important. Whether any of these local companies will reexamine this all-embracing strategy remains hypothetical. Business models will be explored further in Chapter 8.

4.2 Handling the brand

The service provider's brand is a valuable but ethereal asset, as it exists only in the mind of the customer. The brand contains all of the customer's impressions of the service provider's reliability, quality, market positioning, and expected customer service. Service providers have learned to invest in managing their brands with the same commitment as they devote to managing their facilities and business processes.

All telecommunications service providers have to risk the brand to grow, whether through a new service line, a migration to a new technology platform, or a merger with a less-dynamic partner. Without risk, there is no growth, and in some cases, risk is required for simple survival. How service providers undertake these strategic shifts demands skill, and how they handle the inevitable misjudgment sometimes affects the brand more than the original misstep.

AT&T

AT&T threatened its brand with its introduction of WorldNet Internet service. Like other ISPs, AT&T offered free service and software for users to install. Unprepared for the number of interested customers, AT&T ran out of software at its New Jersey headquarters. This did not prevent an unanticipated number of customers from going on-line to test the service. Customers experienced busy signals, unacceptable response times, and an ill-equipped customer service organization. Not long thereafter, a five-hour outage prevented users from collecting their e-mail. AT&T undoubtedly responded by upgrading its system and revising its forecasting models; the problem never recurred.

In this case, the strength of AT&T's brand was the catalyst for the problem. Customers flocked to the service, because the brand signified quality communications. Disappointing users can result in harm to the original brand in addition to the new service. A similar promotional fiasco prompted the U.S. Department of Justice to threaten AOL when it continued to sign new subscribers while service levels plummeted. AT&T, however, had more to lose, because its high-quality reliable network is the foundation of its reputation. The value proposition for AOL is more complex, and after its mandated network upgrades, AOL's service returned to levels acceptable to its customer base.

AT&T's original entry strategy for local markets involved reselling the service of the incumbent local provider. One experiment in consumer local service involved reselling local provider lines at a 22% discount. In four and a half months, AT&T was able to provision only half of the lines it sold. Besides the large revenue loss, the company had to convince the remaining customers to return to their previous carriers. Experiences like this one probably contributed to AT&T's decision to pursue a local cable strategy rather than resale.

MCI

In August 1999, after a software upgrade, MCI's frame relay network failed for approximately 10 days and disrupted service to an estimated 30% of its customers. While outages are disruptive to customers, and network planners need to avoid them to the extent possible, they are inevitable. A year earlier, for example, AT&T had a similar outage for roughly two days. AT&T provided frequent updates to its customers and shared any information it had as soon as it was available. After MCI restored service, customers and MCI itself agreed that there were better ways to handle the failure. AT&T's deft handling of the prior situation created a point of reference against which MCI was measured—and stumbled.

Most of MCI's missteps were tactical errors, exacerbated by the company's inability to pinpoint the problem and communicate it to customers. MCI avoided making public statements in the mistaken belief that an explanation for the failure was imminent. The crucial strategic component is MCI's apparent lack of a brand strategy that includes protecting its image when the unavoidable failure occurs. MCI's solutions included reverting to a previous version of the software, reimbursement to customers, and reviewing its backup networks and its network platform. While the eventual actions are commendable, the fact that a communications plan—and to whatever degree feasible, a restoration plan—was not in place, represented a strategic negligence of the company's valuable brand.

4.3 Opportunity costs

Opportunity costs refer to those events, revenues, or advantages that do not occur because of some alternative path that the service provider chose instead. If a service provider chooses to enter a new market instead of buying back stock, the option not chosen represents an opportunity cost if the new market entry initiative fails to live up to expectations or if the stock increases dramatically compared to the new market's performance. Opportunity costs are often hypothetical, representing a comparison between a known value and an estimate of the potential value of the alternative. Nonetheless, strategic decisions need to recognize that any action or investment precludes other actions that could have superior results. Some service provider failures had less to do with their actions and more to do with what they might have accomplished instead.

Internet fax

Before users were familiar with unified messaging, and before most customers were using e-mail routinely, Internet fax arose as an alternative to paper-based scanning and faxing. Transmitting faxes across the Internet became commercially available in 1997. Internet fax had the potential to save customers about 50% of their fax costs (now Internet faxing has almost no cost at all), yet the technology has not succeeded. Packet-based fax utilizes the network more efficiently than does conventional circuit-switched fax. As with most new technologies, the quality was once inferior to circuit-switched fax, but it improved significantly over time. ISPs have good reasons to sell the technology. According to International Data Corporation, fax transmissions in 1998 constituted an $83 billion market, and the market researcher predicted fax services to grow to $90 billion by 2000 [5]. Even a small percentage of that traffic, at a fraction of the price, diverted to ISPs would represent millions of dollars in incremental revenue. With so much pressure on fixed subscription revenues and ISPs hungry for additional premium services to sell to subscribers, one would expect a significant campaign to introduce the service. Most of the national ISPs had the capability to offer the service, so deployment was not the roadblock. Strategic reasons for its failure include the following:

- *ISPs missed the speed-to-market window for fax before unified messaging became feasible.* Most unified messaging services include the Internet-based fax service and produce the same cost savings or better. Unified messaging, while available through Internet providers, is also available outside of the ISP distribution chain. Missing the technical window for establishing a messaging brand through fax could have damaging effects on the ability of ISPs to differentiate through messaging and collect revenue for incremental services.
- *Rates for international circuit-switched calls fell constantly, reducing the price advantage of Internet faxing.* This is another example of speed to market being critical. If a large price discrepancy enticed customers to use Internet faxing, users would become comfortable with the new technology. If, after service providers reduced the price advantage of Internet fax, customers were already in the habit of faxing over the Internet, they would still be drawn to small discounts or

the potential to eliminate paper from the process. Without the early customer education, the opportunity was lost.

- *ISPs did not develop a marketing plan specifically for the service.* Viewed as a value-added offering to Internet access, ISPs used the same sales force to sell access and faxing. Selling an application takes a different approach from selling access, and the sales prospect for Internet access in a business is probably a different individual than the prospect for fax. Indeed, offices often decentralize the responsibility for fax, although the purchasers of fax machines or telecommunications services are generally easy to identify. Furthermore, Internet fax sales require a different commission plan for sales representatives. Whatever the case in a particular business environment, the sales process for Internet fax probably requires a different strategy from that of Internet access.

ISPs will be able to recover with unified messaging services of their own, but their proprietary opportunity for the market is gone. Unified messaging is available on enterprise communications systems, on hosted sites on the Internet, and by subscription outside of any Internet service provider. If ISPs had exploited the fax opportunity when it became available, they would have had two years to build a brand, to develop new features based on user preferences, and to gain the profits accorded to first-to-market providers.

Holding court

Most of the history of telecommunications was under regulatory supervision. Monopoly providers won change or kept the status quo through courtroom skirmishes with regulators, customers, or competitors. Pundits teased that MCI was a "law firm with an antenna on the roof" for the first decade of its existence, and it was not a coincidence that its corporate headquarters were located in Washington, D.C.

Regulators and the courts are dismantling the vestiges of regulatory rules. Although most telecommunications executives profess that they eagerly anticipate the day that competition reigns and regulation is virtually extinct, telecommunications service providers continue to initiate court action whenever they determine that the level playing field is disadvantageous to their interests. In all fairness, both the former monopolists and the newer entrants actively initiate disputes. More than four years after the Telecommunications Act, industry participants in the United States do not believe that a blueprint for competition has been resolved.

Most of the grievances are legitimate. The strategic error is that the energy expended in their negotiation and resolution needs to balance other competitive inroads. Companies also need to be aware that customers, investors, and competitors are drawing conclusions about each company's competitive readiness by observing which battles they choose to fight.

For example, AT&T has consistently complained to regulators that local Bell companies were not unbundling and reselling their local lines to ensure a competitive market. Its inability to connect satisfactorily with local companies to resell local service was undoubtedly a factor in AT&T's shift to a cable access strategy. This strategy culminated in its acquisitions of TCI and MediaOne and its strategic partnerships with other cable providers. Local governments franchise cable providers, as they do with telecommunications service providers. Like the copper loop, there is only one cable facility in any jurisdiction. After on-the-record complaints that local exchange providers should provide wireline access to competitors, once it was itself a cable provider, AT&T decided that cable systems should be unavailable to resellers. AT&T's reasoning was that sharing its facility would discourage the company to invest in its infrastructure. Eventually AT&T proposed a cable resale arrangement that was more in line with its terms and its timetable. AOL has completed a turnabout in the same direction. AOL lobbied frequently to open cable lines to multiple providers. Since its merger announcement with Time Warner, AOL halted its lobbying activities and espoused a more conservative approach.

RBOCs are just as unwilling to open their networks to competitors. While espousing a wholesale and retail strategy, RBOCs fought to prevent line sharing, which enables CLECs to add DSL services to an existing line. Line sharing reduces the costs for the CLEC by eliminating the need for them to buy an entire line to resell, and it significantly reduces provisioning time, a point of conflict between CLECs and RBOCs. RBOCs jeopardize their own strategic objectives in two ways by continuing this confrontation. First, CLECs can be effective agents to sell the facilities-based services of RBOCs. If at least some of the CLEC sales are incremental to the RBOCs' own efforts, delaying these sales sacrifices revenues. Second, RBOCs exacerbate an already tenuous relationship with their wholesale customers.

While companies are entitled and even obliged to ensure that the law protects their interests, it is not evident that the major providers are striking a balance between courtroom aggression and marketplace assault. The potential opportunity costs of continuing courtroom battles include the following:

- *The protections of monopoly previously limited the consequences of "regulatory lag" to customer frustration; in competition, the price is churn.* Regulatory lag refers to the timeline for resolving regulatory conflicts. Without competition, customers will wait for resolution of regulatory issues when no competitive alternatives are available. In a competitive market, service providers that willingly delay services will simply lose their customers to service providers that do not. The dynamics of competitive technology markets change within months. In open markets, regulatory barriers create makeshift alternatives. By the time regulatory issues are resolved, no matter who is the winner in litigation, the market might no longer be there for the victor to claim the reward.
- *Courtroom delays signal to investors that the telecommunications provider is conservative and risk-averse.* Telecommunications stocks are historically stable and yield predictable dividends, appealing to retirees and risk-averse investors. While telecommunications providers do not want to lose their base of shareholders, they are interested in attracting the investor that their high-technology competitors attract. The new investor looks for growth but not necessarily dividends. These investors offer inexpensive capital to build facilities or make acquisitions. Attracting new investors is behind the trend of creating tracking stocks for higher growth, higher risk divisions. These investors are more aware of the daily activities of companies in their portfolio, including courtroom initiatives.
- *Courtroom delays signal to risk-taking employees that the telecommunications provider is conservative and risk-averse.* Previously regulated companies need to change their internal culture, and one way to do so is to hire staff with values in line with the competitive marketplace. Present employees take cues from management that the company value is to maintain the status quo or win outside the usual playing field. Conscientious employee candidates conduct research before deciding on an employer. These desirable individuals will learn about the competitive activities and the delay tactics of potential employers. A conservative approach to competition will be self-fulfilling.

- *Developing courtroom skills rather than marketplace shrewdness is a competence whose days are numbered.* Taking the human resources allocated to regulatory victories and reassigning them to competitive activities develop their knowledge about the changing marketplace and create skills with a long life cycle. Return on that investment will enable telecommunications providers to drive the competitive market and its associated technology rather than negotiate it.
- *The targets of most regulatory battles are the customers that service providers want to please and retain.* Many of the ongoing legal maneuvers involve the interconnection of telecommunications providers. AT&T's open access controversy concerns the resale of broadband cable facilities. Similarly, line sharing involves resellers of copper access facilities. Most RBOCs have identified the wholesale sector as a market of strategic importance, yet their disputes tend to delay this operating relationship. Courtroom dialogues do not constitute quality customer care.

  The user's price for local service, the primary object of litigation, is well under its current costs. Arguments made on both sides of the issue revolve around companies that want to offer local service but fear large losses. This is a good reason to protect one's position. It would be reassuring to see a similar effort applied to promoting competition through marketplace activity. If local service providers believe that resale charges are below a sound market level, they should be reselling services in other territories and exploiting the arbitrage opportunity. If IXCs seeking to offer local service believe that resale prices are too high, they should build their own facilities. The fact that neither of those outcomes is widespread casts doubt on most litigants' intentions.

None of the previous examples is serious enough to permanently damage a company's overall strategy, reputation, or financial stability. Indeed, strategic blunders cause companies to assess and revise their planning processes to guard against repeating the experience.

References

[1] Shapiro, C., and H. R. Varian, *Information Rules: A Strategic Guide to the Network Economy,* Cambridge, MA: Harvard Business School Press, 1999.

[2] DeVeaux, P., "Internet Gone Global," *America's Network*, Vol. 103, No. 15, p. 32.

[3] Cazzani, S., "Fixed Line Is Superseded by Cellular," *Telecommunications, International Edition,* Vol. 33, No. 10.

[4] Grambs, P., and P. Zerbib, "Iridium Customer Care," *Telephony*, Vol. 235, No. 19.

[5] Gerwig, K., "The Cold, Hard Fax," *tele.com*, Vol. 4, No. 5, p. 31.

5

Rewriting the Rules for Success

Saying there is a bandwidth glut is like saying we had too many transistors in the 1960s.

—Chief technology officer

Developing an appropriate strategy for the telecommunications market takes analysis, intellectual honesty about one's own strengths and weaknesses, and a willingness to challenge conventional wisdom. The years of competition in the few markets that can be called level have already demonstrated that competitive markets work as predicted. On the other hand, competitive markets are shattering some of the conventional wisdom that tends to underscore business plans. Telecommunications providers need to sort through the legends and develop working assumptions that can guide their strategies.

5.1 Porter's competitive strategy model

Michael Porter devised what has become a particularly popular depiction of competitive strategy [1], shown in Figure 5.1. The model, which

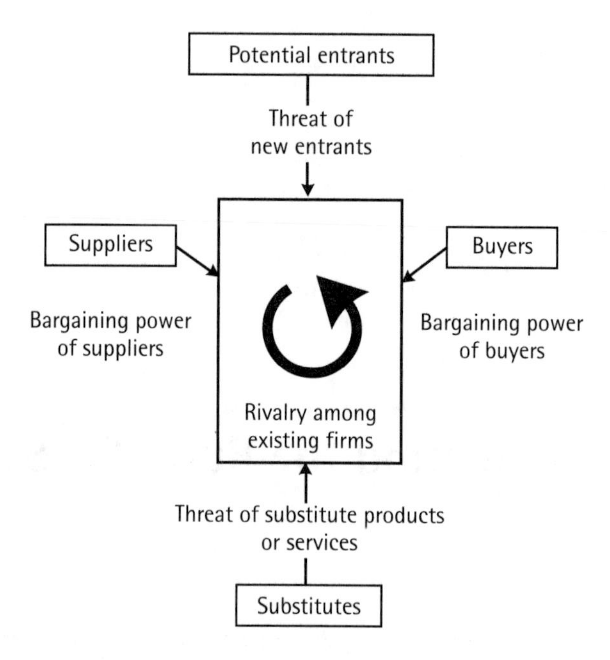

Figure 5.1 Porter model of competitive strategy.

illustrates the dynamics of a competitive market, is useful for describing the nature of competition. Furthermore, Porter's model offers both a common terminology and a set of organizing principles that are critical in a discussion of telecommunications strategy.

The model of five primary forces reflects the importance of market participants outside of the major competitors. Customers, suppliers, and competitors considering market entry affect the rivalry in the marketplace. Among existing competitors, the following factors intensify rivalry:

- *Numerous or equally balanced competitors:* Government regulation, and eventually a mature competitive market, will provide this to service providers. Furthermore, the list of potential competitors increases because of globalization and low barriers to entry for many telecommunications market segments.
- *Slow industry growth:* Participants in mature industries need to fight for share, which intensifies competition. This is not a factor in the telecommunications services industry.

- *High fixed or storage costs:* For facilities-based carriers, fixed costs include the network infrastructure and the cost of licenses. Resellers often have high fixed costs because of volume guarantees to their suppliers. Service providers need to ensure that they cover their costs by increasing market share and revenue per customer.
- *Lack of differentiation or switching costs:* Switching costs are low for customers, and the high level of churn in competitive telecommunications markets demonstrates that many customers view services as undifferentiated. Regulators have been effective at reducing switching costs through local number portability. Differentiating telecommunications services is among the foremost challenges to service providers in the near term.
- *Capacity augmented in large increments:* To maintain a cost advantage, service providers often plan high fixed costs and low variable costs. The consequence is that service providers are under significant pressure to fill unused capacity, intensifying the marketplace. For the largest networks, significant upgrades occur in step functions. A service provider moves quickly from excessive network utilization to high costs and underutilization after the upgrade is completed.
- *Diverse competitors:* Competitors are diverse in size, country of origin, and strategy. Telecommunications service providers face mature competitors from areas that were once unrelated or marginally related to their core transport business, such as local cable providers, ISPs, and entrepreneurs with new access and transport technologies.
- *High strategic stakes:* Decisions such as whether or when to move to an IP architecture have significant consequences. Moving too soon could be exceptionally costly and disconcert existing customers; waiting too long can miss the market.
- *High exit barriers:* For facilities-based providers, there are regulatory barriers to exiting unprofitable markets. Furthermore, abandoning network infrastructure and stranding investment is costly. Last, service providers need to have a large footprint to benefit from economies of scale and scope, so exiting selected markets is not always prudent, even when individual markets are insufficiently profitable.

The threat of entry of new participants depends on the present barriers to entry and the existing competitors' reaction to a new entrant. Barriers to entry include economies of scale, product differentiation, capital requirements, switching costs, access to distribution channels, cost disadvantages not related to scale, and government policy. Barriers to entry are relatively high for new industry entrants but not overwhelming for those already operating in a different region or a complementary technology. Thus, many telecommunications leaders within their own territory have announced plans for global expansion or extension into new markets. The convergence evident in telecommunications services creates an array of substitutes for each service. This trend will escalate as technology advances.

5.2 Conventional wisdom unraveled

Academic knowledge of competitive markets such as Porter's model is valuable, but each industry has its own idiosyncratic principles. Telecommunications service providers need to select the appropriate guidelines from conventional wisdom and their own experience and debunk or adjust the assumptions that might mislead management. Table 5.1 lists several legends and updates them to match the market experiences of telecommunications service providers. The respective sections further develop these legends and describe how emerging assumptions will guide telecommunications strategy.

5.3 Competing against real and strategic competitors

To a telecommunications provider, a *real competitor* is a service provider that offers an equal or substitute service to the same customer segments. A *strategic competitor* can be defined as a company that is well positioned to become a real competitor through territorial growth or vertical integration, or because entering the market creates leverage for its facilities, distribution channels, or customer base. RBOCs and IXCs are strategic competitors for the others' markets. International companies poised to enter new markets are strategic competitors for the existing providers in the market. Usually, strategic competitors arrive with their own set of disadvantages, such as unfavorable economies of scale, conflicts with

TABLE 5.1
Legends and Emerging Assumptions

Section 5.3	Legend: Competitors are easy to identify and difficult to compete against.	Emerging assumption: Major competitors can emerge at a moment's notice, and strategic alliances are blurring the line between competition and partnership.
Section 5.4	Legend: A five-year planning cycle (or longer) is essential for investment-driven businesses.	Emerging assumption: Without losing sight of long-term goals, flexibility is essential in a rapidly changing world; time horizons are shrinking.
Section 5.5	Legend: Investment requirements serve as the major barrier to entry in technology markets.	Emerging assumption: Technology barriers to entry are shrinking, and even resellers can deploy technology to differentiate their offerings.
Section 5.6	Legend: Companies need to get closer to customers to learn what they want and then provide the service.	Emerging assumption: Customers rarely decide what they want until they see the service and use it; they always want lower prices.
Section 5.7	Legend: Markets for telecommunications services eventually become saturated and competition will then intensify.	Emerging assumption: Deregulated telecommunications markets experience intense competition and significant growth, a trend that shows no sign of subsiding.

customers or existing partners, and lack (or age) of infrastructure or diversion of focus.

In a mature and stable market, competitors are easy to identify. In the telecommunications services world of interconnection, partnerships, mergers, and convergence, keeping track of industry participants is a bigger challenge. Deregulation and globalization made possible by the industry itself have complicated matters further.

Few commonly describe Microsoft as a telecommunications service provider. Still, Microsoft has invested more than $100 million each in telecommunications companies Qwest Communications, Teligent, Rogers Communications, and Nextel Communications and $1 billion in Comcast in addition to its own ISPs Microsoft Network (MSN) and WebTV [2].

Competing against strategic competitors includes more variables than conventional competition with real competitors. For one thing, strategic competitors are most often already successful in a complementary market, as they do not yet serve one's existing market. This complicates the dynamics of the competition. AT&T's ownership of a wireline long-distance network enabled the company to redefine wireless pricing when it introduced Digital One Rate service. Companies with established markets can integrate vertically to break even or suffer a small loss if the vertical integration results in a compensating increase in their core business markets. Microsoft created new demand for its Web-enabled operating systems and application software when it invested heavily in businesses (such as WebTV and MSN) that would broaden the customer base for on-line access. ISPs that offer basic personal computers either free or for a very low price threaten the low-end lines of PC manufacturers and the branded Internet services. Bundled computers and Internet access offer a revenue stream of advertising for the package providers. They simply do not need to earn profits on either the Internet access or the hardware. Incumbent providers need to anticipate which companies are likely to enter the market and how they will provide value to the customer.

Telecommunications services draw together couplings of providers that compete in some markets and cooperate in others. In some cases, telecommunications service providers draw the lines at services; in others, the definition is not so clear. In the DSL market, incumbent local carriers have invested in data CLECs at the same time they fight court battles about colocation and other operational and strategic issues.

This solution has become so prevalent in the industry that it has garnered the term "coopetition." Coopetition, a concept credited to Novell founder Ray Noorda, refers to businesses that simultaneously cooperate and compete with each other. Microsoft's investment in Apple to revive the struggling PC maker created a new market for Microsoft's Internet Explorer browser at the same time it strengthened the Macintosh operating system's ability to compete with Windows. The deal also raised eyebrows at the U.S. Justice Department. AltaVista became an Internet portal, not simply a search engine, in competition with the portals that have placed the AltaVista engine on their home pages for their customers to search the Web.

Succeeding through coopetition is not the same as intentionally creating channel conflicts with one's own customers. Coopetition increases sales and profits beyond what they would be without the business partnership between two competitors or strategic competitors. For two reasons, channel conflict results in reduced sales and profits for the business that enters its customer's marketplace. First, the lack of focus for the business in channel conflict cancels out the benefits that vertical integration would provide to increase sales to customers. Second, buyers or potential buyers of its offerings will seek out suppliers that do not sustain channel conflict, or they will create their own products. Hughes Electronics, the leading manufacturer of communications satellites, created its own services subsidiaries, Hughes Communications and DirecTV. Eventually it sold its manufacturing division, eliminating potential channel conflict and creating shareholder value for its owners.

5.4 Keeping up with the changing future

The forces of competition, rapidly advancing communications technology, productivity improvements due to information technologies, and globalization shape the telecommunications industry, and their interaction creates geometric change. Competition accelerates change, because competitors cannot control the pace of service development, price reductions, or innovation. Competitive providers need to introduce products as soon as they are available, even if that erodes their present revenue streams. Competitors pass along cost reductions in the form of lower prices, unless they are certain that their competitors do not have access to the same low costs. Competitors introduce innovation to retain customers in the event that competitors develop similar services.

New communications technology increases the rate of change through its own quickening pace, and because technical development branches out even while it moves forward. While fiber optics developers create new bandwidth algorithms for existing facilities, they also formulate new, higher capacity facilities. At the same time, wireless developers are inventing fixed broadband facilities to compete with the wireline infrastructure, creating a branch to competitive or substitute technologies.

Productivity improvements create cost reductions, and globalization increases the field of competitors and widens opportunities of scale and scope, placing further downward pressure on prices and squeezing time to market. Within this environment, the strategic planning process needs to operate more efficiently than at any other time in history.

Predicting the impact of future technologies is only slightly more sophisticated now than it was in the days of alchemy. Still, the following guidelines help telecommunications providers focus and improve their strategic planning process:

- *Take large financial risks or technology commitments only in the shortest possible time horizon.* In a vertically integrated monopoly, it is possible to control the technology platform, the migration path to a new platform, the timing of deployment, and the timing of phase-out. In a dynamic, competitive market, these options are unavailable. Telecommunications providers will need to choose between stable, long-term technology commitments or having higher costs but more flexibility. When the first electronic telephone switches were installed, estimates for their depreciable life was in decades, because they were built to operate for decades. PC-based systems will operate for decades as well, but their usable life spans approximate their warranties. The smaller the time horizon, the more nimble telecommunications providers will be able to be.
- *Take forecasted changes beyond the technologies to the applications using the technologies.* The type of technology is important, but a proper strategic forecast needs to imagine the uses for the technology and the changes to the customer's operating environment that result from the applications. Until enterprises reengineered their business operations, executives commonly complained that information technology had not delivered on its promises to reduce costs and improve business processes. When the reengineering passion took hold in the business community, its results were palpable and probably contributed to the U.S. economic expansion in the 1990s. It is instructive to look at the "Jetsons" animated television series to imagine a future that simply automates the present without changing business processes. The show's visualization of the future is within a 1950s set of activities. The adolescent son does homework on a computer by turning knobs and pressing buttons, without a digital component in sight. The concepts of the user interface, hyperlinks, or digitalization were absent and, in fact, did not emerge until PC technology was widely in use. Technology companies need to hone their imaginations rather than wait for the marketplace to demand applications. For example, in the early 1990s, a telecommunications service provider told an audience that customers would soon be able to connect their computer to their bank and apply for a loan, implying that computers would simply automate the paper process. Instead, the World Wide Web has overturned the transaction. Customers instead can go to a marketplace, present their needs

and credentials, and select the lowest rate offered by a participating lender.

A more suitable cartoon role model would undoubtedly be Dick Tracy, with his wrist radiotelephone. Besides having accurately foreseen the technology eventually used for wireless services, the cartoon predicted the application of the technology as a solution to a problem—mobility.

Table 5.2 demonstrates the difference between automation advancement and application advancement through a set of examples. Both types of transactions exist; the transformation transactions exploit the uniqueness of the technology.

- *Distrust forecasts with hockey-stick characteristics or categorical statements about markets that do not yet exist.* It is crucial to consider the source of a forecast, especially when it involves markets for which there is not yet a base of customers. Even forecasts that are not self-serving are at best an educated estimate of an unknown outcome. The best approach is to gather as many predictions from as many credible and objective sources as are available. Most forecasts overestimated demand for asynchronous transfer mode (ATM), because many customers stated to market researchers that they intended to adopt the technology but did not. If other technologies become

TABLE 5.2
Automation Versus Transformation

	AUTOMATION-BASED PROGRESSION	APPLICATION TRANSFORMATION
Retail	Eliminate trip to store or bank	Weekly notification of sales
Device functionality	On-line grocery shopping	Wireless telephones locating nearby stores, sales, and services
Information dispersion	Research and education	Interactive games and auctions
Broadcast	On-line publications	On-line communities
Publication sources	Professional	Amateur

available, a new platform comes into view, or price structures among available alternatives change, events can invalidate a technology forecast. Many misguided futurists are still awaiting the paperless office.

- *Review the existing and forecasted state of telecommunications from five years ago before trusting a forecast for five years hence.* To get a perspective on the accuracy of a strategic plan, review trade publications or business plan forecasts from the same time horizon backward as the present view forward. Distance-insensitive long-distance rates appeared in the United States in 1995, at nearly 30 cents per minute. In July 1995, Amazon.com began its on-line operations. The market can change momentously in a few years.

Business planning, like investments, can be more flexible if they are more frequent and less wide-ranging. Telecommunications providers will need to revisit the timing of their business planning cycles. The new telecommunications provider will undergo interim planning cycles between the major strategic initiatives. Strategic and tactical planning will differ from each other more in their targets than in their recurrence. Successful planners will regularly redraw the future.

5.5 Technology as barrier to entry and as differentiator

Deregulation eliminates the most impenetrable barrier to entry, but the telecommunications industry itself imposes its own obstructions. Facilities-based contenders perceive technology to be a most significant barrier to market entry. The investment in today's telecommunications network is immense. New entrants with newer technologies can duplicate equivalent capacity at a much lower cost. Still, facilities-based providers will spend billions on network technology.

There are many strategic reasons to own rather than resell a network, and other incentives to resell rather than own. Chapter 6 covers the issues and analysis to resolve that planning dilemma. Whether or not a telecommunications service provider decides to approach the market through facilities ownership, the provider can maximize the investment either through differentiation or by creating an entry barrier to others. Whether or not a telecommunications service provider is a network owner, there are several ways to minimize competitor opportunities and differentiate service:

- *Compete through newer technologies.* Some of the newest facilities-based providers have built networks designed both for the newest technologies and the target customer markets. Without the FCC's universal service requirement, new carriers can select the most lucrative markets to install fiber optic local and interexchange networks. As the industry matures, as do its underlying technologies, the ownership of a network is less of an entry barrier than it once was, and this trend will probably continue. Qwest uses dense wavelength division multiplexing (DWDM), which enables the company to build its infrastructure at a lower cost than its competitors do with legacy systems. When the average cost to build a T3 circuit was approximately $1,400 per circuit mile, Qwest's cost for the same circuit in its initial buildout was less than $500 [3].
- *The provider can offer proprietary features.* This approach requires either an internal R&D capability or contractual commitments from outsourced developers. Proprietary features do not need to reside in the network, which makes this differentiator available to facilities owners and resellers. Programmable add-on switches can offer the benefits of proprietary technology without the commitment of network ownership. Programmable switches use application programs to control call sequences, redundantly to the public switched network. Skills in innovation rather than investment amount will make the distinction between the successful and the undifferentiated. Recognizing the value and flexibility of programmable switching, switch manufacturers such as Lucent and Nortel are moving toward open architectures.
- *Providers can differentiate their services through information technologies within other business processes.* If providers choose to offer unadorned commodity services to their target markets, or if providers can only create network services that are equivalent to those offered by their competitors, information technology can provide a technology differentiator outside of the network. Already, competition is simmering in areas such as data warehousing, on-line provisioning, and customer care. Again, the telecommunications service provider that innovates in addition to its technology investment can

find differentiation opportunities that competitors cannot easily imitate.

5.6 Customer buying patterns

Customer desires are often difficult to predict. Market surveys proliferate, but for new services, the data they yield are not rigorous enough to build a business plan, let alone a network. A featureless dial-tone service with one service provider per territory can estimate both usage and revenues with little more market data than a census and a building forecast. New services, especially digital services, will render customer surveys less useful than in the past.

Surveys often provide excellent information about buyers, but they are unable to capture the subtleties of purchasing behavior. Survey after survey concludes that customers are eager to bundle services. Occasionally, though, a more specific survey states, as categorically, that customers resist commitment to a single provider, that a customer needs a large discount to consider bundling, or that customers generally do not want to bundle certain services. Bundling is easily comprehensible, because it includes only services the customer already uses. When surveying customers about new services, such as the previous example of ATM deployment, respondents can only estimate their future behavior. Technology can change between the time of the survey and the deployment of the service. Prices can be higher or lower than anticipated, which will change everything if the service demand is highly elastic. A substitute technology can arrive or fail to arrive when expected, decreasing or increasing usage. Outside factors such as a sweeping change to the economy or changes to a business customer's ability to invest in the short term can invalidate survey results. The explosion of Internet usage created a largely unanticipated demand for second lines for consumers. Several RBOCs and GTE suffered provisioning slowdowns in their efforts to meet the demand.

Customer demands are insatiable for both service and price. It is tempting to state that customers' stated desires are unreasonable, because they want excellent service and very low prices. Nevertheless, there are three reasons why these demands are quite reasonable. First, customers have had high-quality service from most telecommunications providers for as long as they can remember. Asking reliability to continue—or for new entrants to match current performance—is rational. Second, prices have fallen in telecommunications and virtually every technology business. Expecting prices to stay low or go lower is simply the expectation that the market will stay the same. Third, the combination of very low prices and very high service might indeed be impossible to obtain from a single provider. If customers cannot find the combination they desire, if they need the service, they will buy some combination that is available in the market. Customers will be willing to sacrifice quality or pay a higher price than they would like or live without the service at all. If, though, they discover a service provider with the service and pricing that they seek, they will buy. The challenge for service providers is to ensure that competitors cannot create a disparity between their price/service offerings and one's own.

Finally, the diversity and complex capabilities of emerging technologies will complicate customer buying patterns. It is impossible to sell new services in a sound bite. For a telecommunications service provider hoping to increase usage, simple measures, such as promotional pricing or mainstream advertising, will suffice. For providers selling a broadband service, or a software-driven Web-hosted package, or a feature that has never existed before, customers require education and then persuasion. The failures of ISDN and the various all-in-one network solutions offered by IXCs are examples of the service provider's inability so far to demonstrate the benefits of new services using traditional media such as television advertising or bill inserts.

5.7 Usage accelerators

Several characteristics of the industry have accelerated the demand for telecommunications services. To the degree that service providers can control the following factors in a cost-effective manner, they can increase their own growth rates:

- *Lower prices:* One unexpected result of deregulation is the considerable elasticity of demand for telecommunications services. Elasticity occurs when price reductions cause an increase in demand for a service. In highly competitive markets such as those for long-distance and wireless service, demand is growing at rates that almost exceed the price reductions. While much of the supporting data is anecdotal, the increase in usage for telecommunications services that follows significant price reductions offers a compelling argument that usage will increase when prices fall. This trend is certain to continue. Local telecommunications prices are slowly approaching both the level and structure of their costs. IXCs have begun to charge a fixed charge for monthly service in an effort to match their prices with their cost profiles. The most auspicious trend, though, is that the cost of bandwidth is dropping, and bandwidth providers show no sign of limiting their expansion programs.

- *More bandwidth:* Facilities-based providers are building bandwidth energetically. The traditional backbone providers in the United States—AT&T, MCI WorldCom, and Sprint—are bolstering their digital backbone networks. Newer providers such as Williams Communications, Level 3, and Qwest have built similar networks or announced plans to do so. The ready availability of capital, coupled with the predicted demand for new bandwidth to handle traffic growth and broadband applications, have fostered a race to build nationwide facilities. In Europe, British Telecom, Cable & Wireless, Global Crossing, GTS, and a partnership of KPN and Qwest are building backbone networks. Other partnerships and lone U.S. carriers plan to build networks to capitalize on the estimated 100–200% growth in demand for bandwidth.

  Technologies such as DWDM promise to increase the bandwidth on fiber backbone networks. The technology multiplies the capacity of individual optical fiber strands and reduces costs by a factor of up to 10. DWDM is in use in intercity backbone networks and locally in metropolitan area networks (MANs). In the United States alone, DWDM networks plan to increase capacity by more than 2,000 times in just a few years. According to a Yankee Group forecast, carriers will build as many as six new networks between major European cities, each with potential capacity geometrically higher than current networks [4]. A study by Ovum Research states that carriers intend to direct more than half of future spending toward DWDM equipment with 16 or more channels [5]. Yankee Group and other market researchers have predicted that the unit cost of backbone bandwidth will approach zero in the near future.

- *Pricing innovations:* Several structural paradigms in wireless pricing have proven to accelerate usage. The near flat-rate pricing simplicity of AT&T's wireless Digital One Rate and the other services copying its pricing plan boosted wireless usage more than sheer elasticity would predict. Bell Atlantic Mobility experienced a 100% increase in sales when it introduced its single-rate plan. Long-distance flat-rate per-minute pricing is responsible for some of the growth in long-distance calling. Flat monthly rates for unlimited Internet access with unlimited local calling have propelled the United States to become a world leader in Internet subscriptions. There are network planning considerations to offering unlimited plans, especially when not all costs are fixed. The original launch of unlimited monthly Internet access plans degraded AOL service sufficiently to require the intervention of the U.S. government. Since several well-publicized debacles, service providers have been cautious when introducing unlimited plans to encourage users to consume under-used facilities, such as weekend long-distance calling.

The apparent simplicity of long-distance charges is causing some customers to use the service inefficiently, presenting an unintentional revenue boost to service providers. Most low per-minute plans include a monthly service charge. Customers who do not use a minimal number of minutes would have a lower overall bill if they subscribed to a higher per-minute rate that carried no service charge. Other plans offer the low rate only during off-peak hours. Customers inadvertently using the long-distance network during the business day can incur rates four to five times the rate they expect. Still, the ease with which customers now make long-distance calls, compared to only decades ago, demonstrates the cultural change caused by price reductions and simpler pricing models.

Wireless pricing philosophy CPP is in use outside the United States and is under consideration within the United States. CPP requires the caller to a wireless number to pay the charge for the call. In some ways, CPP is a counterpart to removing roaming charges. Roaming charges discouraged customers from using their mobile telephones outside of their home territories. The absence of CPP discourages customers from leaving their telephones powered on, even in their home areas, for fear that they will be forced to pay for calls they do not want. With CPP, customers can feel free to leave their telephones on and give their mobile numbers to others. Some wireless providers without CPP have provided the incentive to leave the telephone on by offering the first incoming minute at no charge. Customers can then continue calls for which they are willing to pay.

CPP also helps customers control their telecommunications

costs, as they can make only the calls for which they are willing to pay. This benefit becomes more important as wireless penetration begins to include the cost-conscious consumer market. Where it is in effect, CPP has increased wireless usage by 25–50% [6]. Outside the United States, where customers often pay by the minute for wireline local service, CPP helps wireless telecommunications become a threat to landline communications. As wireless per-minute prices fall, they approach landline prices and thus become a competitive alternative to a second line or even a first line.

Differences in the industry structure between the United States and countries where CPP is successful could hinder its introduction in the United States. Some countries use a special prefix for wireless numbers that informs callers that the call will be billable. Service providers introduced CPP in most non-U.S. markets when wireless penetration was quite low, avoiding an education process that the United States would need to bear. The Strategis Group has found that 75% of U.S. callers do not want to pay for calls to wireless telephones.

- *Reduction of switching costs:* Another regulatory initiative with the potential to increase competitive activity, lower prices, and increase usage is local number portability (LNP). LNP is one of the FCC checklist requirements imposed on the RBOCs before they can offer long-distance services in their own territories, and it was a European Union directive for implementation, along with carrier preselection, in January 2000. LNP will increase competitive activity because a telephone number change is a significant switching cost for both consumers and businesses. Consumers dislike the inconvenience of notifying contacts of a new number. For businesses switching providers, changing telephone numbers imposes out-of-pocket costs for stationery and collateral materials. They also stand to lose customers who are unable to locate the new telephone number. LNP enables customers to switch carriers effortlessly and will undoubtedly be a primary driver of competition and churn in local markets.

- *Flexible terms:* Wireless providers in the past amortized the high price of the wireless handset in the monthly subscription rate. This was necessary to attract customers who were not willing to pay hundreds or nearly a thousand dollars for their first wireless telephone. Now, however, prices have fallen; customers are more confident of their need for a wireless telephone; and new market entrants have created downward pressure on prices. It is no longer necessary to bundle the high price of a handset into the contract price for most customers. This has resulted in the elimination or near elimination of the service contract, which was once necessary for the cellular provider to recover the costs of the customer's hardware.

- *Deregulation:* While service providers cannot accelerate the pace of deregulation, they can impede it. Deregulation increases usage because new entrants create supply, which places downward pressure on prices, because a multiplicity of providers encourages manufacturers to innovate and reduce costs, and because competition tends to diversify the array of service offerings. To create more utilization in exchange for more risk, telecommunications service providers can set a priority to eliminate regulatory barriers in the industry.

References

[1] Porter, M. E., *Competitive Strategy: Techniques for Analyzing Industries and Competitors,* New York, NY: Simon & Schuster, 1998.

[2] Christopher, A., "Enemy Mine," *Telephony*, Vol. 237, No. 22, p. 22.

[3] Noel, J., and M. Langner, "The Stock Market," *tele.com*, Vol. 2, No. 11.

[4] McClure, B., "Europe Gets Backbone," *Telecommunications, International Edition,* Vol. 33, No. 6.

[5] Wright, K., "Tomorrow's Network Made Simple," *Telephony*, Vol. 237, No. 13, p. 50.

[6] Kridel, T., "CPP Now or Never?" *Wireless Review*, Vol. 15, No. 15, p. 36.

6 Strategic Decisions

Customers want both high quality and low price, and that is just impossible.

—RBOC product manager

Before a telecommunications service provider can open its doors for business, it faces several strategic decisions. The resolution of these deliberations will determine the investment strategies, marketing plans, operations principles, and overall direction of the business; in short; they affect every aspect of the company. Management needs to address each of the alternatives and combine their answers into an overall strategic direction.

6.1 Quality versus price debate

Telecommunications service providers confront a marketplace where customer expectations are high and competition is fierce. As shown in Figure 6.1, the conflict between service quality and price is inevitable.

It is apparent that a telecommunications service provider cannot sustain a high level of capacity in comparison to its traffic level and maintain

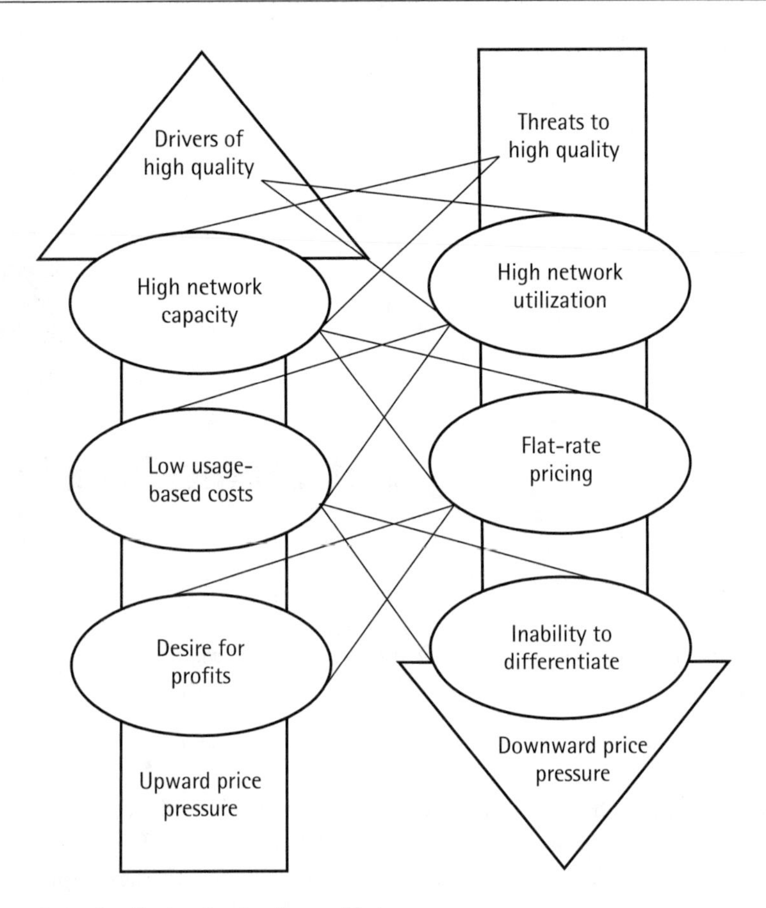

Figure 6.1 Quality and price in conflict.

the high utilization (the opposite of excess capacity) required to keep costs low. The trend toward flat-rate pricing—not per-minute pricing, which is usage-based—will undoubtedly lead eventually to unlimited monthly service for some customers. Several IP-based providers already offer this price structure, and Sprint is experimenting with unlimited usage plans in the consumer long-distance market. As AOL learned in the Internet market, unlimited service increases usage. In Europe, where free Internet service is more common than in the United States, usage charges for local calls mitigate the potential overuse of unlimited Internet plans. The arrival of unmetered calling will undoubtedly increase both Internet penetration and usage in Europe where it is available.

Compounding the challenge of quality and price tradeoffs, the service provider's present inability to differentiate appreciably based on quality is at odds with its desire to maximize profits. One reason that differentiation is so difficult for providers is that few service providers claim control of the entire communications experience. IXCs and local providers jointly offer service. Business customers that connect directly to long-distance providers often maintain their own internal networks. Despite the publicity for single-source providers such as AT&T's Integrated Network Connection, Sprint's ION, and MCI WorldCom's On-Net service, few customers actually receive end-to-end, accountable network quality. The Concert partnership between AT&T and British Telecom intends to reduce some of the multiple-provider mosaic of international provisioning and service. Even this massive enterprise needs to interconnect locally in areas not served with its own facilities.

There are other forces exacerbating the difficulties of maintaining a high level of service quality. Service providers need to confront these realities, described as follows, directly through their strategic planning processes:

- *Providing the highest quality service is axiomatic for incumbent providers in developed regions, and anything less could threaten the brand.* Monopolists offer a high-quality, if homogeneous, service to customers. Furthermore, telecommunications service was relatively featureless until quite recently. Maintaining reliability for narrowband service was quite manageable for service providers and ensured the silence of customer groups and regulatory commissions, at least about network service quality. Thus, customers who have experienced telecommunications services for decades expect and receive reliability beyond almost any other utility and beyond any other appliance. Customers rewarded the incumbent providers with superior brand strength, so service provider investments in the network delivered an important return. Nevertheless, excellent quality is most often unrecognized, and less than excellent quality is intolerable. Whatever the customer's expectations, the high overall network quality did not come without cost. Telecommunications providers, especially those with circuit-switched networks (still the only alternative for the highest quality voice service) support high-cost profiles.

- *Perceptibly lower quality commands price differences beyond the difference in cost for known technologies.* As a high quality is the baseline expectation, new entrants to the marketplace have discovered that very low prices are required to lure customers to a lower quality alternative. IP telephony presently falls in the category of a low-cost service with a perceptible difference in quality. Many variables will narrow the gap in the near future. First, quality of IP-based telephony is improving, and its use is becoming more transparent to the customer. Early systems required computers and proprietary software between both participants in the call, for example. The proof that IP is becoming more mainstream is that incumbent providers, local and long-distance, are testing IP architectures; manufacturers are developing new technologies and migration paths; and industry analysts agree that existing circuit-switched networks will migrate to IP. Second, costs for IP networks continue to drop. Third, IP networks promise to add features at a lower cost than traditional networks and potentially add features that cannot operate on a circuit-switched network. Whether the lower costs and lower rates will result in higher economic profits will require research on the part of the individual service provider.

- *Telecommunications service providers that choose to differentiate through network quality will need to maintain their own networks.* In all probability, the telecommunications service provider that commits to a quality-of-service differentiation strategy will need to obtain proprietary access to the network and its differentiating quality. As a rule, service providers that choose this strategy will operate their own networks. For providers that currently own networks, this is a viable strategy. For providers that do not own networks but would like to differentiate on service quality, it is still a viable strategy. A modified strategy for telecommunications service providers without large networks would require proprietary contracts with R&D firms, contracts with wholesalers with strict service level agreements, and network enhancements such as ancillary software to support exclusive enhanced services. This approach can be especially useful in niche markets, such as vertical markets, especially if bandwidth becomes a commodity and the service provider finds more productive uses for capital than in building conventional network facilities.

A service provider can consider several actions to resolve the quality-of-service dilemma. First, a facilities-based provider or one with ready access to network operations can configure the network to prioritize traffic based on its type. For example, delay in a database transaction will be much more obvious to the on-line user than an equal delay in the delivery of an electronic message or fax. Emerging policy-based network management tools are becoming available for managing quality of service and security based on user-applied parameters.

A second solution is to segment customers and the service quality they require if the opportunity is available. The service provider should identify those customers with a need for superior service and a willingness to pay—business customers, for example. Then, if possible, the service provider can separate them from those who are willing to receive downgraded service in exchange for lower prices, most likely certain consumer segments. Besides widening the array of service packages, which will please customers seeking variety, segmentation can offer increased revenues compared to the traditional method of providing averaged services at averaged prices. The service provider can accomplish this through a low-price subsidiary; or, if the lower-price service threatens the core brand, the service provider can license or sell its services itself through unaffiliated brands.

Third, the service provider can improve its price structure to ensure that usage-driven costs drive measured prices. This will be challenging in highly competitive markets where flat-rate services have become popular. Nevertheless, the telecommunications service provider that can match its cost profile to its price structure will be able to maintain the high quality demanded by the customers who are willing to pay for it. Occasionally, a telecommunications service provider finds that it cannot offer the quality that a customer needs at the customer's willingness to pay. If that occurs, either the customer needs to evaluate his or her demands (because they are impossible to meet) or the provider needs to evaluate its cost structure (because competitors can meet the customer service needs and price requirement).

Last, the telecommunications service provider can achieve brand equity based on quality attributes outside of conventional network service. This strategy is available to facilities-based providers and resellers. With an acceptable but not exceptional level of transmission quality, the telecommunications service provider can differentiate on other network-based services, such as enhanced call management services, messaging services, or new services presently unimagined. This strategy is particularly appealing to providers that seek to command a brand premium but that are generally unable to differentiate their network quality through traditional measures such as reliability or signal quality.

6.2 Build versus buy decisions

The decision to build and maintain a network—or some of its components—is complicated. There are other make-versus-buy decisions that are as strategic but perhaps not as momentous. Conventional wisdom once contended that resale is merely an entry strategy. The notion was that companies would enter markets through resale and gain critical mass, then build and maintain their own facilities. Actual experience in the U.S. long-distance market, which is a relatively mature competitive market, has challenged that assumption. According to Atlantic-ACM, about 43% of inter-exchange service providers are facilities-based, while more than 38% are switchless resellers.

In the local consumer market, the decision to build or buy is especially problematic, because prices for residential local services are historically below economic cost. In other words, wireline local networks lose money in the consumer market. This presents a disincentive to build. If the in-place network is run efficiently yet sold below its cost, then it stands to reason that a new network, built at similar cost, will not achieve profitability in the consumer market. Many assumptions reside in that hypothesis, though, so the variables deserve further exploration. Furthermore, emerging local wireless standards such as Bluetooth and fixed wireless services will complicate the cost paradigm for service providers. There are reasons for building a network, and corresponding reasons to buy instead. AT&T, MCI, and an assortment of other providers intended to enter the local market through resale, but they opted to build facilities instead. The decision rests on the service provider's position in the marketplace, competitive strengths, and alternatives available in the markets the provider plans to serve. The tradeoffs involved in the decision to build or buy local networks are described as follows:

- *Profit potential:* In general, owning facilities and providing service are cheaper and more profitable than reselling the facilities of another provider. Not all consumer markets maintain the same cost profile.

Serving residential customers in densely populated areas can generate returns of 20% [1]. Returns in the local business market could be higher yet. Technology markets do not always conform to the rules of other industries characterized by high fixed costs. New technologies and falling costs could create overcapacity in the local market, which will have a more significant impact than in the transport market. Backbone networks in the interexchange market are sharable. Network owners with too much capacity can sell unused minutes on whatever circuit is available at the desired time. This flexibility is not available to local facilities providers. The copper or fiber optic or wireless link between a network switch or node and the customer is a dedicated facility. The rooftop antenna, in-ground copper wire, or other facility that duplicates the customer's purchased facility is simply stranded, unless it can be moved to a willing buyer. Unless the provider can find an alternative use for the facility, no revenues will be realizable.

- *Competitive advantage:* Generally, local resellers view incumbent local providers as impediments in provisioning services. Several contenders for local competition, including the largest IXCs and aggressive CLECs such as USN Communications, McLeod USA, and e.spire Communications Inc., have begun to pull back from resale strategies to concentrate on network buildout [2]. This attitude, if adopted by many potential competitors, bodes poorly for competitive penetration into high-cost residential markets in suburban and rural areas. The discount applied to resold facilities tends to average out the attractiveness of entering markets through resale. If buildout costs are high, then resale discounts harm incumbents and resellers in a similar manner. The establishment of unbundling network elements—components of service that incumbent service providers must sell separately to resellers—enables competitors to choose the elements to resell and choose which elements to offer on their own. In fact, a strategy of unbundling and some resale enables service providers to maintain only networks that offer them competitive advantage, which is an argument in favor of resale over buildout.
- *Control of market:* Building one's own facilities enables the service provider to control pricing, services offered, and service quality. Facilities-based carriers view this control as an opportunity to get closer to the customer. On the other hand, resellers argue that

applying those resources to marketing and support enables the reseller to be closer to the customer. Purchase also enables service providers to enter or exit markets when profitable opportunities are present. Network buildout is a long-term commitment to a particular market. Resale offers flexibility, a benefit in a highly competitive marketplace.

- *Proprietary information systems support:* Building enables the service provider to develop proprietary information systems, such as OSS systems that work to differentiate the service from that of its competitors. On the other hand, avoiding the commitment of in-house systems enables the provider to avoid purchasing or developing, maintaining, and updating legacy information systems. In addition, as OSS systems migrate towards industry standards, their opportunity for significant differentiation shrinks. Last, the service provider needs to assess the cost of maintaining its own support against the importance its proprietary system holds to its customer base.

- *Control of business resources:* Resellers were critical of WorldCom's proposed acquisition of Sprint, because the resulting industry consolidation would have reduced an already small field of wholesale network providers. In the past, resellers of long-distance service could choose between MCI WorldCom and Sprint, among other providers. Emerging network providers, including wholesalers, are rushing in to fill this gap. Still, this reseller concern underscores the vulnerability of resellers that have to depend on their suppliers for most of the service they provide. On the other hand, buying can free up capital, an important resource, because resellers can generally purchase their inputs on a just-in-time basis. Facilities-based providers need to invest in the network in advance of the sale. Resellers can better match their costs to their revenue streams.

- *Control of technology costs:* Again, this decision point is a dual-edged sword. Building enables the service provider to amortize its costs over a long time and therefore stabilizes future costs. Conversely, buying creates flexibility to move to better or cheaper technologies as they become available. If the technology itself is a service differentiator and network ownership is a strategic necessity, then the added control is a benefit.

- *Simplifying cost management:* Building networks assists service providers in developing cost management skills. They learn the costs and capacities of network equipment, and their cost profiles improve as they move from one market to the next. Resellers have to manage an assortment of prices from all jurisdictions. On the other hand, resellers can gain profits by managing their costs carefully between the discounted resale cost and the selling price to customers.

- *Building company value:* Establishing a facilities-based network creates a valuable asset for investors or acquirers needing to expand into new markets, so builders have an advantage. Ordinarily, though, networks do not fit together like puzzle pieces, and acquirers need to sell duplicate portions of the acquired network in an unfavorable market. Resale creates an attractive mass of customers for acquirers who already own networks.

Other opportunities to outsource abound. Service providers, even those that have made the decision to own networks, are quite willing to outsource other aspects of operations, such as OSS systems, billing, or receivables management. Similar strategic questions are present in making any outsource decision. Can the company differentiate and brand its service because of this superior in-house function? If not, then outsourcing is a viable option. Does in-house management of the function reduce costs substantially during its entire lifetime? Outsourcing information technology management or software development is feasible, especially when system upgrades are included in the contract and transparent to users. If the functionality of the systems involved does not strategically differentiate the service provider from its competitors, the service provider can choose to outsource the systems.

Even when the functionality of systems and other in-house talents do differentiate the service provider from its competitors, it can still be an outsourcing candidate if the benefit of the differentiation is a small part of the overall business. For example, suppose a telecommunications service provider had an exceptional trouble-reporting-and-repair system, well ahead of those of its competitors. The strategic question is whether to maintain the proprietary advantage of the system and incur all of its costs. However, excellence in trouble reporting and repair, while an integral part of the telecommunications provider's value proposition to the customer, should not be a frequent enough customer experience to affect the customer's choice of provider. Indeed, if customers are quite familiar with trouble ticketing, the service provider probably has deeper problems. This presents a dilemma for the service provider: whether to maintain a system that could well be slightly more costly than those supported by competitors or purchase the function from an outside source and give up the competitive advantage. The entrepreneurial provider will be able to do both, by divesting the system developer completely. The telecommunications service provider can then purchase the business system from its divested offspring and share the costs with the new customers of the developer, its competitors. If the developer begins to penetrate markets not presently in competition with the service provider, all the better, but that is not necessary. A true divestiture will eliminate the channel conflicts that the service provider would have if it tried to market the systems to its competitors while it still owned the operation. The large benefit of lower cost will offset the small competitive advantage that the service provider would have as the proprietary operator of the system.

6.3 Economies of scale and scope

Economies of scale occur when businesses can conduct their activities most economically with large investments in fixed assets. In a fundamental business, break-even analysis, a large fixed asset base results in a profit schedule in which the highest profits occur at high sales volumes. A business operation with no fixed costs and all variable costs has the same percentage of profit on each sale or transaction. Most businesses, of course, have some fixed costs and some usage-based costs. Telecommunications service providers have historically sustained a cost profile of high fixed costs, funded with a high level of debt. Monopolies, characterized by low business risk, can sustain higher-than-average debt successfully. The interest rates that monopolists pay for borrowed funds is more cost-effective than borrowing through equity.

Businesses achieve economies of scope when they share business resources to serve additional markets. The shared resources can be physical assets, costs across the value chain, information, or external relationships. For example, as local and long-distance services use the same distribution network, obvious economies of scope accrue to the service provider that offers both services to the same customer. Besides the improved utilization of the plant, the service bundle can share the monthly bill, distribution channels, and customer information. Because regulators somewhat arbitrarily separated local and long-distance services from each other, there are better examples of economies of scope in the industry. Internet access and portals offer economies of scope. Cable television has created economies of scope with Internet access. Any vertical integration opportunity is likely to present economies of scope.

Few would dispute that early turn-of-the-twentieth-century telecommunications networks were in all likelihood a "natural monopoly." Customers regarded telephone service as a luxury, and indeed it was, because telecommunications was a routine part of neither business nor community. Local networks used costly equipment and were labor-intensive to build and maintain. Once they became monopolies, telecommunications service providers took steps to ensure that their networks were cost-effective and that charges were predictable. The absence of competitors enabled providers to control the pace of new technology, offer very standardized services, and keep prices low. These are good arguments for maintaining a large fixed asset base.

Whether the network continues to be a natural monopoly is now a matter of some controversy. Whether economies of scale still govern the business is also uncertain. The centralized architecture of the traditional network is giving way to smaller switches and a more distributed architecture. Switching was never the sole source of economies, and some economies of scale and scope are present. Still, service providers choosing to remain small can do so profitably, if the decision is part of a larger strategy.

There are many reasons to attest that economies of scale are waning, or at least becoming more complex. Moore's Law, named for the founder of Intel, predicted that computer chips would double in power every 18 months. After repeated confirmations, the prediction still holds true for the computer industry. Some network equipment costs (for routers and small switches) are falling at rates characteristic of other computers, even though large switches are more costly and more capable, and software and other labor-related costs are rising. Outshining Moore's Law, the amount of information that can travel over a strand of fiber doubles every 9 to 12 months.

Like computers, network devices demonstrate scalability—the first purchase does not have to involve a giant monetary commitment. Serving the first customer in a market does not require a multimillion dollar investment. After the first subscriber in a market, costs for new entrants are similar to those of incumbents. In the reality of the telecommunications market, incumbents have huge scale but their networks are old. New entrants do not have the scale, but they benefit from the new technologies that are less expensive to purchase, operate more efficiently, and require less labor to maintain. Software, like equipment, has become scalable. Enterprise resource planning software, once out of the reach of small and mid-sized companies, will be available on a transaction basis through application service providers, accessible on the Internet. These packages offer the same functionality of the purchase-only versions. Not all systems are available in lower scale portions, so new distribution techniques do not eliminate economies of scale. They simply diminish them. In addition, new entrants benefit in software in the same way they benefit from new hardware. The telecommunications service providers with large-scale systems are generally maintaining legacy applications at a higher cost than new software technologies.

Scale can be a disadvantage. The obsolescence conundrum affecting facilities-based carriers multiplies by the scale of the network. A commitment to a wide-scale technology needs to last for the forecasted period. Scale limits a telecommunications service provider's flexibility to transform to meet market requirements, including technology independence and the ability to enter and exit markets to maximize profits. Economies achieved from scale often reduce the service provider's ability to offer specialized services, a necessity in a competitive market. Another trend will emerge when markets operate freely; telecommunications service providers will be able to select the customers that they choose to serve. Monopolists do not have the option of withholding service from undesirable customers, but competitors have some flexibility in that area. Obtaining critical mass was once a primary requirement for growth; as the market matures, service providers will be more interested in the quality of customers acquired through their sales efforts or mergers than in quantity alone. The concepts of mass or critical mass will recede in importance to the concept of lifetime customer value. Scale by itself is not a prerequisite.

Information businesses maintain significant economies of scope, but they do not require economies of scale. Traditionally, companies attained economies of scope by improving the utilization of fixed assets. In the information economy, many marginal costs approach zero. After a document or rentable software program is on-line, the sale of the ten thousandth copy produced no more cost than the first copy did other than distribution costs, which are negligible in proportion to the cost of development. Information providers and software developers create economies of scope by distributing their products at very low cost in new markets. These economies do not demonstrably contribute to the utilization of fixed assets.

Nevertheless, there are several well-founded arguments that telecommunications service provision sustains economies of scale and scope. Executives advocating their newest merger announcement most often extol the economies that they expect to achieve. While these predictions sometimes serve as publicity, the mergers between regional and national providers will indeed promote the additional buildout of long-haul networks. None of the seven divested RBOCs in their regional contours could justify building a nationwide fiber optic network, as most of the network would serve territories far from their customer base. Nonetheless, the new service areas of SBC and Bell Atlantic/GTE now have a nationwide look—if a bit spotty. These mergers will create the feasibility of nationwide network buildout and whatever economies they offer. Mergers involving underserved regions such as Latin America and Asia will benefit from industry consolidation, with or without the investment of overseas service providers. When areas are costly to serve, as are those with low population or old infrastructure, they can improve at a faster rate when their parent companies generate investment funds from other, more lucrative ventures.

There is no question that the success of AT&T's One Rate programs, both wireline and wireless, would have been impossible without network scale economies. IXCs could not offer domestic flat-rate calling without either nationwide networks or contracts with nationwide suppliers.

Other economies do not relate to networks but are still valid economies of scale. These include nationwide advertising programs, the ability of network providers to standardize and control all aspects of their service, and a large size to improve a service provider's access to capital. Large providers can serve multinational corporations without creating ad hoc partnerships and shared account responsibility. Regulators should note that high-cost local provider US West improves its chance to survive by joining a multinational lucrative enterprise like Qwest, where its unprofitable rural networks make up a smaller percentage of the corporate portfolio.

Branding creates additional economies of scale. In certain markets, strong brands can afford to create new service providers simply for the sake of additional branding opportunities. Retailers frequently launch their own prepaid long-distance cards, either for promotional attention or to ensure that customers see the brand name whether or not they are near the store or the store's advertising. The United Kingdom's Virgin has extended its own brand to the telecommunications industry, with offerings in both wireline and wireless markets. The retailer's intention is apparently to extend its brand to new and quite visible markets. Virgin's success in branding its airline (with its huge logo painted on its airplanes offering valued recognition) created a viable corporate enterprise, while its cola brand was not successful. Convergence offers economies of scope. Telecommunications service provider and cable provider entry into the Internet access market, alliances and mergers between service providers and content providers, and the market entry of electric and pipeline utilities into the telecommunications transport market illustrate this weekly.

6.4 Setting market share goals

Market share is a company's total sales as a percentage of total sales in a defined industry. All businesses need to set market share goals in their strategic planning process. Whether a particular market share goal is realistic depends greatly on the structural characteristics of the market the service provider chooses to enter.

Monopolists have 100% market share in that they serve the universe of possible customers in their designated geographical and product line areas. Monopolists, usually in partnership with regulators, control price levels, price structure, and other defining features of the market. The presence of competition converts monopolies into other market structures, and consequently many other market characteristics change as well. Markets can exhibit four structures: monopoly, perfect competition, oligopoly, and monopolistic competition.

A market in *perfect competition* has so many competitors that no one seller can influence pricing. In perfect competition, all competitive products are equivalent, customers have access to many sellers, and customers are well informed about their competitive choices. The Internet services market in the United States is an example of perfect competition. Because the services are relatively undifferentiated, customers switch providers with ease, creating churn.

Only three or four competitors control most of the output in industries with *oligopoly* as its structure. Therefore, oligopoly most resembles monopoly, in that each market leader has enough market share to influence industry attributes. Some industries are oligopolies because of licensing, as is the wireless market. Others are oligopolies because entry barriers are high, such as the facilities-based interexchange market. As the entry barriers (that is, the cost of building a network) fall, more companies enter the market. The facilities-based interexchange market is moving to a fuller competition with the entry of new network-based service providers. Other oligopoly markets will become more competitive as new licenses for fixed and mobile wireless technologies begin to compete with wireline and wireless incumbents. As new entrants become successful in a market, the original market leaders cannot control prices as they did in the oligopoly structure.

In *monopolistic competition*, there are many market participants, and market shares for each are relatively small, at least small enough that any one competitor cannot influence the overall level of price. Nonetheless, monopolistic competitors can influence the prices in their immediate and narrow niche. For example, a long-distance reseller, among thousands of resellers, cannot set a price that differs greatly from those of its competitors, because the reseller market is in perfect competition. The reseller can choose to redefine its market position to operate within a different market structure by differentiating the product to meet the needs of a targeted segment. By doing so, the reseller can create a monopolistic advantage in the submarket of a larger marketplace in perfect competition. A service provider that specializes in small business can devise a service package that its customers view as unique. The service provider can then enjoy the benefits of monopolistic competition, including the ability to command a premium price.

No telecommunications provider can control whether the market it has chosen to serve will be an oligopoly or a market in perfect or monopolistic competition. All providers need to set market share goals, and successful service providers will set goals that place them among the leaders in their defined marketplace, however the business plan defines the segment. Companies with a relatively small share within their delineated market risk becoming an acquisition target by the leaders in the category.

Companies monitoring their progress against their business plan objectives will pay close attention to market size and market share. If a service provider's revenues grow at 25% in a market that is growing at 40%, the service provider is losing market share to competitors. Managers that focus primarily on sales growth are likely to overlook these early indicators.

On the other hand, telecommunications service providers that overstress market share relative to other performance measures can inadvertently take actions that diminish profitability. Paying a large premium for an acquisition merely to increase market share can be foolish if market share is already at an appropriate leadership level. Initiating price wars in the hope that low prices will increase one's share—in an oligopoly, where prices tend to be inelastic—generally results in lower profits for all of the providers participating in the fray. Leading telecommunications service providers in oligopoly markets need to set market share goals that are realistic within the market structure and in line with one's own competitive strengths. Moreover, market share goals and their subsequent achievement need to be moderate enough not to invite regulatory scrutiny. Domestic companies with high market share are often candidates for antitrust actions. Market share that satisfies local regulators can concern those overseas. U.S. companies seek European Union approval for mergers that could affect EU member markets.

Providers that are dissatisfied with their success in markets of perfect competition do have the option of changing their market focus or segmentation strategy to move to a market of monopolistic competition. As the telecommunications market is still rather uncharted, it is not obvious what constitutes an ideal market share. It has required a mind-set shift for former monopolists to recognize that market leadership involves plurality in some markets and majority in others.

The benefit of market leadership depends upon the product category. Some emerging research suggests that in some markets expansions of market share actually diminish the perceived quality of a brand [3]. The effect is not as pronounced for commodity products as it is for well-differentiated products. Service providers whose business model depends on an image of high quality need to take into account the effects of increases in market share on their brands. Fortunately for providers, premium pricing tends to mitigate the negative perceptions of brands caused by market share gains.

While it is possible to recover from errors in strategic decisions, any miscalculations in these large issues are most difficult to reverse. Most of the decisions are at the core of the business design. Many, such as the build-versus-buy decision, require a significant investment with a long-term payout. Telecommunications service providers need to consider the implications of each of these decisions in the context of their targeted markets, their financial circumstances, and the specific business goals of the enterprise.

References

[1] Philmon, E. W., R. P. Kissell, and C. Janson, "Why CLECs Should Build Their Own," *America's Network*, Vol. 101, No. 23, p. 12.

[2] Branson, K., "Is Local Resale a Sinking Ship?" *Phone+*, Vol. 13, No. 6, pp. 52–54, 56, 58, 60.

[3] Hellofs, L. L., and R. Jacobson, "Market Share and Customers' Perceptions of Quality," *Journal of Marketing*, Vol. 63, No. 1, p. 16.

7

Strategies for Success

Execution IS strategy.

—RBOC chairman and CEO

Businesses widely regard strategic planning as crucial to organizational success, but until recently, the findings from many companies' strategic planning processes were nearly too broad to be quantifiable. Consequently, many planning efforts resulted in schemes that required interpretation before implementation. Technology has changed every aspect of business, and strategic development is no exception. Telecommunications service providers conduct strategic planning efforts with more depth, more frequency, and more involvement throughout and outside the organization.

This chapter serves as an introduction to the rest of the book. The book's structure divides strategy development into four domains: planning, marketing, operations, and finance. Each domain introduces several paths to differentiation as a starting point for telecommunications providers.

7.1 Misconceptions about strategy development

Misconceptions abound concerning the development and implementation of strategy. Perhaps at one time only senior management developed strategy, annually and in smoke-filled rooms, and handed the resulting instructions to the working masses for rote implementation. If that were ever a depiction of the strategic planning process, it is certainly true no more. Strategy suffuses all successful organizations. The strategic planning process is a competitive weapon, and telecommunications service providers need to conduct their own strategic evaluations continuously.

The strategic planning process is valuable even if the planning assumptions are wrong, if the planned profits do not accrue as expected, and if all of the market dynamics change. In fact, many good plans do not identify all of the eventual corrections that will take place during and after implementation. The alternative to a plan that needs constant adjustment is no plan at all. The lack of a plan (or indifference to one) results in no assumptions (with a guarantee of inaccuracy) and uncoordinated and aimless execution. Plans and their subsequent refinements are necessary for success.

Another misconception about strategy that is taking longer to disappear is that senior management can base strategies primarily on their own visions and desires. A valid strategy first needs to incorporate the demands of the market and the promise of the enterprise. At the outset of the strategic planning process, not every telecommunications service provider possesses the resources that it will need to compete successfully, and its ability to acquire those resources before market entry is a necessity. The strategy's implementation plan must include both the current capabilities of the enterprise and any required new capabilities that are within reach. A fine strategy cannot require capabilities that will not be present during the planning horizon. This is a major reason that senior management does not enjoy a free hand in devising strategy. Sometimes this fact is not apparent from the declared strategies of providers. For example, each of the largest U.S. providers has announced for several years that its strategy is to offer a "full solution" or "one-stop shopping," when in fact, none was positioned or equipped to accomplish that until recently. A telecommunications service provider with an unexceptional strategy but the resources to meet it will exceed the performance of the company with the dazzling strategy without the capacity behind it.

A most dangerous misconception about strategy is the belief that strategy is somehow distinct from implementation. Most strategies fail to achieve their objectives. One reason for this failure is that the strategic planning process neglects to itemize the metrics confirming that the enterprise met its objectives. Alternatively, some strategies fail to assign each objective to a manager with the resources and the accountability to return with success, or they fail to test routinely whether the strategic requirements are completed or in progress. Strategy is pointless without a strong tie to its implementation and measurement of its results. The dusty strategic plan binder on a bookshelf is regrettably a standard ornament in the management office.

7.2 Strategy is ubiquitous

Strategy is becoming more engaged in all business operations for many reasons, most, if not all, of which are due to the enabling functions of information technology; while strategy surrounds and pervades the entire organization, information technologies support and permeate the organization's business processes. Several reasons for the growing importance of strategy are described as follows:

- *New tools enable strategists to perform analyses that are more sophisticated.* The advancement of PC technology coupled with the parallel development of sophisticated statistical and modeling techniques has permitted strategic planners to simulate, model, and test strategic planning possibilities with much more assurance than at any time in the past. Data warehouses supply an extensive body of customer information from which strategic planners, marketers, and operations staff can draw. Rigorous analysis tools can help to generate ideas and create hypotheses for further testing. Moreover, communications tools and intranets make it possible to link strategic planners instantaneously with operating managers, regardless of their physical location. This offers an opportunity for an interactive strategic planning process, which quickens its pace and improves the accuracy and viability of the resulting scenarios.

- *New measurement tools enable managers to evaluate and directly relate results to strategic decisions.* After strategies become responsibilities, and plans to actions, improved measurement and analysis tools are available to capture the contributory relationship between strategies and ultimate outcome. This enables operations management to respond to market variations or the success or failure of service offerings to minimize customer loss. Enterprise-wide planning applications now make up-to-date information, gleaned from a variety of company information systems, available for analysis by both line and staff management.

- *Organizations have decentralized the strategic planning function, putting it closer to operations.* Many organizations have moved a portion of the strategic planning function into operating divisions. One reason is to ensure the accountability of line management by delegating planning resources. Another advantage is that divisional management can be more focused on the trends and opportunities facing them and can identify narrower and more focused market opportunities that the parent company planners can overlook. Centralized planners can then concentrate on the opportunities that go across divisional boundaries or that are outside the present scope of the organization. This structural modification to decentralize has its costs. For one, strategic plans developed within the division have the potential to improve an individual division's performance but ignore activities creating greater benefits that accrue to other divisions and appear in another division's profits. Conscientious development of incentives and performance measurements at the divisional level can alleviate this concern. Additionally, decentralizing the strategic planning function potentially increases the number of individuals assigned to the function, and potentially complicates the organizational chart by creating dotted-line reporting relationships that can be difficult to manage. These challenges require examination, but they should not define where the planning functions reside, because they are secondary to the needs of the business. For example, companies dependent on innovation and conglomerates whose businesses are diverse benefit from a broader, less centralized strategic planning function. Telecommunications service providers often fall into one or both of these categories, so they are likely candidates for a decentralized strategic planning function. Some incumbents maintain cultural obstacles to dispersing the planning function, but they will be at a competitive disadvantage to those that allow the business requirements to drive the organizational framework.

- *Technology and other market variables have reduced the lag between strategy and execution and between execution and outcome.* The shrinking time window creates a need to revisit the strategic planning process frequently. Especially in technology businesses or industries undergoing deregulation, companies can face the predicament of needing to turn around completely when a market requirement becomes evident. As one example, Microsoft centered its entire operation on the Internet in a matter of months, and well before most other technology companies understood the requirement to be Web-enabled. Whether Microsoft created the Web economy through its vast influence or simply predicted its arrival is less important than the fact that Microsoft completed its strategic planning reassessment and execution within months. Similarly, the transformation of networks to accommodate the forecasted explosion of data traffic will occur in a remarkably short time. Operations personnel need constant involvement in the planning process and its implementation and follow-up.

- *The recent emphasis on customer focus has moved the strategic planning function out closer to the marketplace.* Many companies interact with important customers and suppliers during the development of their strategies. Strategic decisions often result in large expenditures or long-term commitments, and it is necessary to test them to the degree practical before implementation. The knowledge residing on the front lines is invaluable to manage the risks of large strategic decisions. Telecommunications service providers must analyze the information gleaned from these interviews with the understanding that customer and supplier objectives do not always intersect with the objectives of the enterprise. Also, suppliers can be coy about their plans, and customers can honestly misrepresent their needs and purchase criteria. Still, the information gleaned from external stakeholders can be very useful.

7.3 Strategies and the paths to differentiation

One of the foremost concerns for the telecommunications service provider is to cultivate a market position that differentiates its services from those of its competitors. Service providers are among the highest advertisers of any company, and most enjoy substantial company name recognition. Brand recognition is another matter, and customers making purchase decisions consider both the company's reputation and the service under evaluation. Telecommunications services remain a near-commodity and are destined to remain so until customers absorb the network capacity now under development. Competing on a price basis is normal for commodity products, but most telecommunications service providers are seeking alternative competitive arenas to minimize their dependency on price competition. Differentiation of the product is the means to commanding higher prices and consequently higher profitability.

Many excellent theoretical models for strategic planning are available, and most are as pertinent to the telecommunications services market as they are to any other market. Rather than describing in depth management literature that is widely available, this section will apply specific strategic planning issues to the telecommunications services market and identify differentiation opportunities for providers.

Figure 7.1 depicts the four primary domains of service provider strategies. Within these categories are potential differentiation paths that are available to the service provider. In niche markets, there are undoubtedly other opportunities to differentiate. For each path, even before an initial market entry, and continuously before each business planning development cycle, it is vital for telecommunications service providers to assess their current capabilities, the demands of the market, and the gap between them.

The four main strategic areas are planning, marketing, operations, and finance. Information systems technology, now integral to every business function, is present in all of these strategic centers and can support a telecommunications provider's initiatives in every strategic matter. This chapter will define each of the four fundamental strategic centers; the book's remaining chapters will examine the strategic opportunities for success within each path to differentiation.

The strategic planning process within each of these domains should include at least the following:

- A statement of the enterprise's vision and fundamental values as it relates to the domain of planning, marketing, operations, or finance;
- Evaluation of the service provider's existing capabilities and those that are within reach;

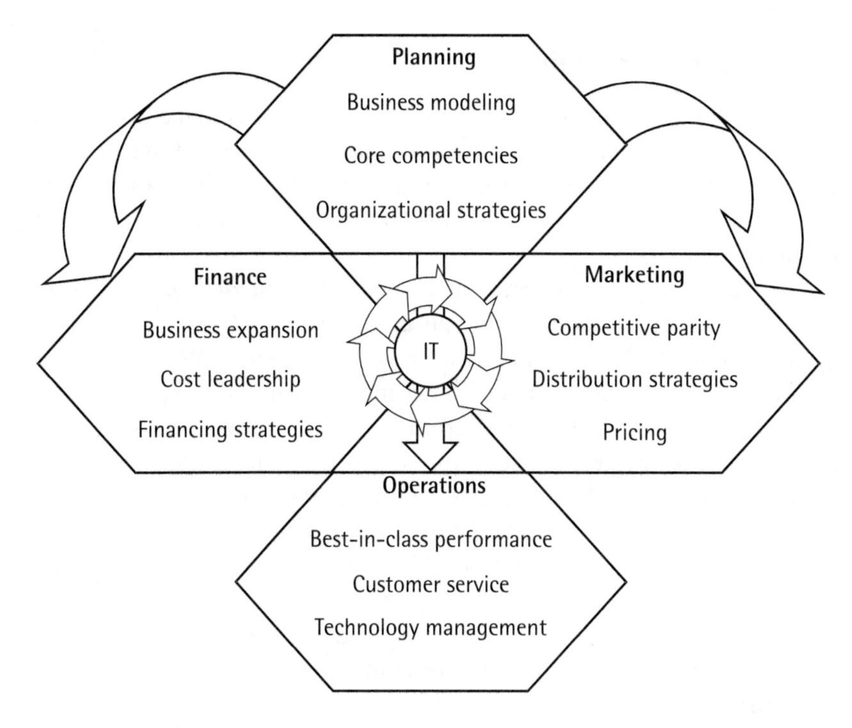

Figure 7.1 Strategies for success.

- A critical competitive assessment of strengths and weaknesses relative to competitors;
- An assessment of the most attractive markets, in view of the markets' growth, competitive dynamics, and the service provider's competitive strengths;
- Estimations of the future direction and growth of selected markets.

At the end of this evaluation, the service provider will hold enough information to decide which paths, if any, offer a competitive opportunity to differentiate services and gain market share. Should the service provider decide to pursue a differentiation strategy, planning tools, coupled with information technology, will assist the service provider in setting objectives, identifying required actions, establishing time frames, and defining metrics to recognize whether the objectives were accomplished.

7.4 Planning strategies

Business planning comprises the earliest set of decisions that a telecommunications provider will make. The planning process encompasses the business goals, investor and stakeholder requirements, and human resources issues within the enterprise and considers the environment surrounding the enterprise as well.

The planning function encompasses *business modeling*, an objective assessment of *core competencies*, and *organizational strategies*. These differentiation paths are among the most fundamental decisions a telecommunications service provider must make before venturing into business or before transforming into a competitive enterprise. The decisions in this phase not only affect the subsequent business choices but are likely to require a significant cultural change among management and employees.

Business modeling involves the development of a business concept. A proper business concept will attract those customers that seek the characteristics in which the telecommunications service provider demonstrates strength. Ideally, the business concept should limit the market to exclude those customers whose purchase criteria include characteristics where the service provider is weak.

Core competencies refer to the process in which the telecommunications service provider constantly reinforces its competitive advantages. The process is a balance of resources; core competencies require investing in strategic differentiators requiring excellence, outsourcing or shaving down functions that require only adequacy, and eliminating those functions that are not required in the targeted markets.

Organizational strategies refer to the enabling techniques of an enterprise to ensure that it uses its resources most effectively to meet the organization's objectives. A telecommunications service provider can differentiate its services through its effective business processes, its efficient organizational structure, or its maximization of in-house talent.

7.5 Marketing strategies

The marketing component addresses the differentiation paths of *competitive parity*, *distribution strategies*, and *pricing*. Marketing is the weakest capability of many incumbent providers and has opened a gateway for new entrants to gain market share. The conventional marketing mix contains the elements of product, packaging, price, and distribution. In an industry in which both the product and the packaging continue to be relative commodities, the distribution strategy and the pricing offer the best potential for differentiation. Skills in managing competitive parity enable telecommunications service providers to maintain an acceptable level of profit and minimize customer churn in an industry characterized by intense price competition and accelerating speed to market.

Competitive parity skills notify the telecommunications service provider when to take a significant market initiative. Sometimes it is appropriate to lead in the marketplace, and sometimes a competitor's action requires a competitive response. For example, Sprint takes an initiative each time the service provider reduces its per-minute price of service. When MCI WorldCom matches the price reduction, it does so trusting that its net revenue loss is smaller at the lower rate than it would be at the previous rate. As telecommunications usage has proven to be elastic, lower prices result in more minutes of customer usage. Because customers are price-sensitive, a customer would be inclined to switch carriers if a matching decrease was not forthcoming from its current service provider. Sprint's own original price-reduction initiative demonstrates the use of competitive parity, because the original price change anticipates the reaction of other competitors.

Distribution strategies enable customers to locate the services they seek and enable telecommunications service providers to locate and serve customers. Distribution channels begin at the supplier, traverse the telecommunications provider, and continue outward through in-house sales channels, agents, resellers, brokers, and retailers. The value of a given distribution channel depends on a variety of factors. These include the proficiency of the participants within the channel, the proper choice to connect with the targeted segment, and the customer's access to product information and alternative channels.

Pricing innovation is essential in the telecommunications services market. First, the services are near commodities, making customers very price-sensitive. Second, even in view of an explosive demand, the short-term threat of overcapacity is real, putting downward pressure on prices. Technology products tend to fall in price, and rising usage is crucial to any service provider whose investors need to see continual revenue growth. Last, new telecommunications service providers need to find profitable pricing structures to overcome historical pricing anomalies, such as below-cost services, complex pricing equations, and settlements between cooperating providers.

7.6 Operations strategies

Long viewed as the workhorse of telecommunications services, operational excellence is becoming noteworthy as a distinct competitive strategy. Customer lack of tolerance for much variation in the quality of telecommunications services creates a marketplace in which all service providers are good performers. Superior quality is difficult to achieve but commands a loyal customer base and premium prices when providers offer telecommunications services with reliability. The differentiation paths of *best-in-class performance, customer service,* and *technology management* represent opportunities within the operations domain.

Many incumbent telecommunications service providers aspire to best-in-class performance because it melds well with their traditional corporate values. New entrants also view best-in-class performance as a potential differentiator in a crowded marketplace. This path to differentiation has many challenges, especially in the area of cost. While performance is undoubtedly important to customers, the standard is already high; prominence in the field is not easy. Furthermore, the areas in which best-of-class performance is noticeable generally require significant investment to participate competitively.

Many telecommunications service providers have stated their missions to differentiate based on customer service. This is an obvious path to differentiation for incumbent local providers, because they are well aware that commodity markets compete through low cost. Most incumbents do not yet support the cost profiles to position themselves as the low-cost providers in their territories, so they are seeking opportunities to command higher prices for superior service. Incumbents also appear to believe that their reputations and infrastructures create a branding opportunity, and this is borne out by customer survey data.

Another way to differentiate through operations is to provide network services of perceptible superiority, and several telecommunications service providers are already mobilizing to take this approach. To accomplish this, service providers will need to be particularly competent at technology management. Technology management can include proprietary innovations, but it can also differentiate through exceptional supplier management.

7.7 Financial strategies

Financial strategies are common to most industries as an opportunity to gain competitive differentiation, and they are just beginning to emerge within the telecommunications industry. The objective of financial strategies is to maximize profitability and growth. As with marketing, the finance strategy will require considerable attention and could require a cultural transformation. Regulated companies structured their financial strategies to meet the goals of regulators and risk-averse stockholders. Competitive markets will require a reassessment of financial goals and the development of new skills in management and differentiation. The paths to differentiation include *business expansion, cost leadership,* and the use of innovative *funding strategies.*

The options for business expansion include territorial expansion, mergers or startups to create additional scale or scope of business, vertical or horizontal integration, or outside investment for the sake of diversification in a cyclical industry. The rapid pace of industry consolidation demonstrates that service providers believe that the benefits of huge scale offset the management and integration challenges they bring. Investors require growth, and they judge the success of a company's expansion somewhat on its own growth and partly on its relative growth while the industry consolidates around them.

Managing the revenue side of the income statement is popular for many reasons, especially in a booming economy, but some telecommunications service providers will succeed through their management of the cost side of the statement. Especially in a commodity marketplace, keeping costs (and the resultant prices) as low as possible is a competitive advantage. The way to win a price war is to avoid sustaining profit casualties.

Funding strategies have changed as much as any other aspect of the deregulated market, but they can be barely visible in favor of more glamorous business processes. Still, the service provider that creates innovative funding and financial management can outpace the provider with higher gross margins and casual fiscal practices.

Each differentiation path comprises both a business process and a plan of action. For example, the business modeling process assesses the market opportunities and the skills residing in or available to the telecommunications provider and develops a sustainable value proposition on which to define the business. The plan of action contains the results of the business process. Part of the business modeling process is to determine which of the remaining differentiation paths is a priority for the chosen business model. Not all differentiation paths are required as a priority, although competitive service providers cannot ignore competitive movements in any of them. In fact, because resources are always limited, reducing the emphasis on selected differentiation paths without damaging the resulting business model is one of the challenges of the process.

7.8 Strategy tools

Strategic planning is more effective and more accessible when the process includes tools. Fundamentally, the purpose of strategic planning is to manage uncertainty. Tools help to accomplish this by forecasting future performance based on past results, estimating the impact of known changes to the marketplace, and speculating about the impact of unknown marketplace changes, such as technology advances, economic dynamics, or competitor actions. Tools can facilitate the decision process for each of the paths to differentiation.

Like other management trends, the tools that companies use to assist them in strategy change over time. Conventional tools such as the evaluation of strengths, weaknesses, opportunities, and threats (SWOT) and the consideration of political, economic, social, and technological (PEST) factors are valuable to develop the strategic overview. Newer tools require more detailed source data and make use of the processing power of PCs and software.

According to strategic planning consultant Bain & Company, the number of tools used remains high, although the most popular tools change within a few years. For example, benchmarking, pay-for-performance, core competencies, and customer retention tools increased in popularity between 1993 and 1998. In the same period, there was a reduction in the usage of customer satisfaction measurement, total quality management, portfolio analysis, and scenario planning. While there is no correlation between any specific tool or the number of tools and improved financial performance, Bain supported the conclusion that there is a correlation between a company's financial success and the way it uses tools. Many factors help to determine whether an organization will be satisfied with its tools, including the fitness of the tool to its application, the readiness of the organization, and the reasonableness of its expectations.

Using tools for strategic planning is effective, but planners need to balance the framework of a particular tool against the need for the process to be creative. Many tools enable innovation, but some primarily increase operational efficiency. A strategic planning tool is most advantageous when it identifies three potential types of improvements: doing existing functions more effectively, eliminating unnecessary functions, and identifying opportunities that are not apparent in everyday operations.

Successful telecommunications service providers will perform everything capably and some functions exceptionally. Telecommunications service providers that choose to serve niche markets can succeed through excellence along only one or two paths to differentiation—if those areas are very important to customers and if performance in the other differentiation paths is satisfactory. Chapters 8–19 describe how to achieve excellence in the areas that become strategic priorities. Each chapter also describes some of the management tools available to help strategic planners develop an understanding of their present market situation and to evaluate their strategic and tactical alternatives.

8

Business Modeling

Bill Gates thinks bandwidth should be free. We think software should be free.

—Network CEO

The rise of interest in the concept of business models is a favorable one; it demonstrates that telecommunications service providers recognize the need to focus. Business models enable organizations to concentrate on the most important and competitive elements they oversee.

8.1 The value proposition

Simply stated, a business model depicts the way the enterprise will offer value to customers profitably. The term *value proposition* refers to the combination of service, price, and packaging or bundling, delivery channels, service quality, customer support, and any other interaction between the customer and the service or the service provider. The element of value is critical, as it embodies the concept that customers are willing to trade up to a given amount of expenditure for the services offered. The value proposition, then, is the proposed exchange of resources that the service provider offers the customer.

The business model defines four essential qualities:

- The unique competitiveness of the service offering;
- The benefit the customer receives;
- The specific role of the provider;
- The profit potential that would make the business opportunity a worthy investment for a stockholder or an entrepreneur.

Each of these elements is strategically indispensable, and the service provider has a great deal of flexibility in defining every one. Without a uniquely competitive service, in a competitive market, the business model will fail. A commodity service can be uniquely competitive, although the product itself is indistinguishable from its competitors, if other elements of the value proposition distinguish it from the competition. One method of achieving competitiveness is through low price. Telecommunications service providers that select a low-price business model need to maintain a superior low-cost profile. Promotions and price wars are undesirable but sometimes necessary in the short term; maintaining a low-cost structure enables telecommunications service providers to offer low prices continually and achieve acceptable returns.

The benefit the customer receives describes why a customer is willing to pay for a particular service rather than using a substitute or doing without it entirely. This is an added hurdle beyond having a unique offering. Many unique products fail in the marketplace, because they are unable to match the product to a customer's willingness to pay. Technology products and services are especially vulnerable when they first emerge. Prices for new technologies often reflect high production costs, the amortization of R&D expenditures, and an estimate of the small market of early adopters.

The business model needs to define the role of the provider to ensure that the service provider has the skills to compete effectively in the selected market. Moreover, defining the provider's role helps to narrow the scope of the business. The scope of the business delineates its boundaries. While most enterprises should position themselves to integrate horizontally, vertically, or outside their present scope to meet changing market conditions, some business models fail because they are simply too broad. Defining a business scope that maximizes the enterprise's strengths—and excludes breadth that demonstrates no apparent strategic advantage—does not limit the potential for business growth. The intensity of competition will provide enough challenge to the market participants without introducing challenges of scope such as channel conflict, unmanageable geographical expansion, or an unfocused portfolio of services. The dimension also includes the characteristics of the distribution model, such as the mix of real and virtual storefronts, the nature of fulfillment, and the positioning of customer care.

Last, the business model needs to demonstrate profit potential to investors and other stakeholders. Whether the funding source is debt, venture capital, equity, or simply the concentrated efforts of an entrepreneur, the business model needs to demonstrate its superiority over alternative uses for capital and other resources. A business model that meets a market need but simply cannot be profitable within the range of a customer's willingness to pay is a failure.

Figure 8.1 illustrates only a sampling of the decisions required in the business modeling process. A telecommunications service provider can choose to serve broad or narrow markets, but a decision to serve one type of market effectively will influence other components of its business model. The same holds true with channel selection and the basis of differentiation. Deregulation causes fragmentation and specialization in

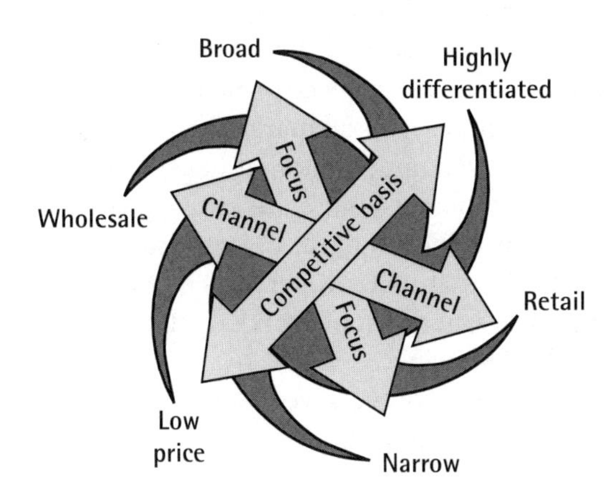

Figure 8.1 Strategic modeling decisions.

markets. This will result in the need for service providers to target niche markets if their goal is to compete by differentiating their services. Alternatively, the service provider that selects the wholesale market as its preferred channel strategy will more likely compete on low price rather than the uniqueness of its service.

Another component of the modeling process requires the telecommunications service provider to abandon traditional strategic thinking to redefine customer needs rather than focus on achieving competitive advantage. Researchers have discovered that companies that generate high growth utilize a strategy called *value innovation,* which makes competition irrelevant by offering new and superior buyer value in existing markets and creating new value to create new markets [1]. Companies practicing value innovation do not use the competition as their strategic reference. The high achievers excelled even during industry downturns by expanding the market and capturing significant share. Value innovation can be independent of technological innovation and is most often independent of simply increasing value within the present service line. Examples of value innovation in the telecommunications services industry include AOL's user interface; AT&T's Family One Rate plan, which multiplied the size of the consumer wireless market; and the Internet auction site eBay, which redefined the retail marketplace.

8.2 Transforming the business model

Telecommunications service providers that move from monopoly markets to competitive markets often neglect to review their business model. A monopolist, by definition, serves all markets in its designated jurisdiction. Monopolists do not face competitors that have developed superior business models in limited markets, but competitive former monopolists will. While it is physically possible for a telecommunications service provider to retain a business model that strives to serve all markets, it is unlikely to succeed in the long term, when competitors target specific market segments with more specialized offerings.

Business models for competitive companies differ from those of monopolists in the following ways:

- *The monopolist mandate is to serve all customers; competitive service providers can select the target market that matches its skills and its profitability requirements.* In the U.S. telecommunications market, 80% of an RBOC's cash flow emanates from only 20% of its customers [2]. Competitive providers are most interested in the top 20% of customers, as is evident from competitive entry in cities and in business markets around the world. Former monopolists (and new entrants, once regulators have ascertained that the marketplace is finally competitive) will have a limited responsibility to provide service in unprofitable markets. Notwithstanding those requirements, telecommunications service providers will have the luxury of selecting their target markets, and their preferences will undoubtedly include the most profitable segments as well. Competition drives prices and margins down. Today's business customer and high-end long-distance user is accustomed to paying rates that subsidize other users. Once prices are closer to costs for all customers, the margins in these currently attractive markets will be quite thin. Other markets that are presently less desirable, such as rural areas, could potentially command higher margins than the conventional business and high-volume markets. While current prices for local service are below the cost of providing service, new technologies are less dependent on density for cost-effectiveness. It is possible to envision a future in which access providers will be able to offer uniform local service pricing in all areas, as they already do for long-distance service.

- *The monopolist offers standardized, uniform services to keep costs low; homogeneous services are simply a building block for competitive enterprises to create value.* Forming a business model requires an understanding of the organization's objectives. When its objectives change, the model will vary. A regulated company holds affordable universal service as its goal and achieves that goal by creating consistent, reliable service. Economies of scale historically served to minimize cost, achieving near-total penetration in most countries and resulting in equivalent service for all customers. Technology has reduced the cost of customizing service, and deregulation has unleashed customer desires for specialization. Customers with a choice of providers will select the service best suited for their own requirements—if differentiated services are available. The differentiation will probably take place in the retail segment of the distribution chain. When the telecommunications services industry divides into its channel components, wholesalers will exploit network economies of scale to offer bandwidth that is plentiful, uniform, and very inexpensive. As the bandwidth moves toward the customer, retailers and other packagers will modify and enhance services to meet the specific needs of targeted users.

- *The monopolist vertically integrates to minimize cost; the competitor builds or buys based on its strengths, its alternative uses of capital, and its need for flexibility.* AT&T's divestiture of Lucent demonstrates one example of the downfall of vertical integration in a competitive market. As a monopoly, AT&T controlled its technology from R&D to manufacture to installation at the customer location. This model kept costs relatively low and provided maximum control of all the elements of service and a small profit on each leg of the journey. Competition encourages companies to participate in the value chain at the point that offers them the best use of their organizational strengths and the highest profit potential, that is, the optimal business model. Even for those companies that intend to offer end-to-end service to their customers, inhabiting the entire value chain is unnecessary. The business model for the competitive industry, for most providers, will include some ownership and some outsourcing.
- *Technology and infrastructure drive the monopolist's business model; customers drive the competitive business model.* To ensure that rates remained low, network planners let the asset-intensive infrastructure drive the business model. An electronic switch that could last 20 years was in service and on a depreciation schedule for 20 years. This is the low-cost alternative to upgrading technology as new systems become available. The result was a business model featuring low prices and slow introductions of new technology. In a competitive market, the switch that can work reliably for 20 years will undoubtedly be technologically obsolete in much less time. R&D costs are higher, because they are duplicated among competitors and because competitors are racing against each other. However, low prices are not driving the business model; customer demands are, and customers apparently want new, faster, feature-rich communications networks.
- *Deregulation fosters innovation, and innovation creates new business models.* The increase in development activity that characterizes competitive markets also innovates at a faster rate than vertically

integrated monopolies. Innovation creates new business models partly because it increases the variety of services available. The variety results both from the emergence of software-defined networking, in which services can be customized easily, and from the willingness of competitors to develop products and services specifically for niche markets.

8.3 New models

A variety of new business models arose when industry entrants assessed market opportunities. Figure 8.2 depicts only some of the business models currently in use by telecommunications providers.

In Figure 8.2, the models vary across one continuum defining their location on the network and across another describing whether the uniqueness of the model depends more on bandwidth, applications, or content. Organizing the variety of business models in this manner demonstrates a host of opportunities for telecommunications service providers of any size. It also demonstrates expansion opportunities for those providers that are eager to move into new markets with a minimum of fragmentation.

The horizontal axis in Figure 8.2 describes the location of the service on the network. Regulators, especially within the United States, in their desire to deregulate the industry in steps, have artificially emphasized the distinction between access and transport. Some physical distinction between access and transport still exists, especially in circuit-switched hierarchical networks. The most apparent partition between local and long-distance service is visible to customers who remain unable to use one provider for both. Another indication of the distinction in the United States is that local calls are most often unlimited without additional costs, while charging for long-distance calls is usage-based. The movement toward measured calling for all calls, and the associated removal of long-distance charges, is blurring the line for users of some wireless services, as are some IP-based unlimited calling packages. Eventually, the concept of long-distance calling as separate from any other communications could disappear entirely. When IP networks are universal, they will further erode any residual differences between access and transport from the customer's viewpoint. Above the bottom of Figure 8.2 (the most hardware-intensive area), the network locations are less strict but most likely to occur either

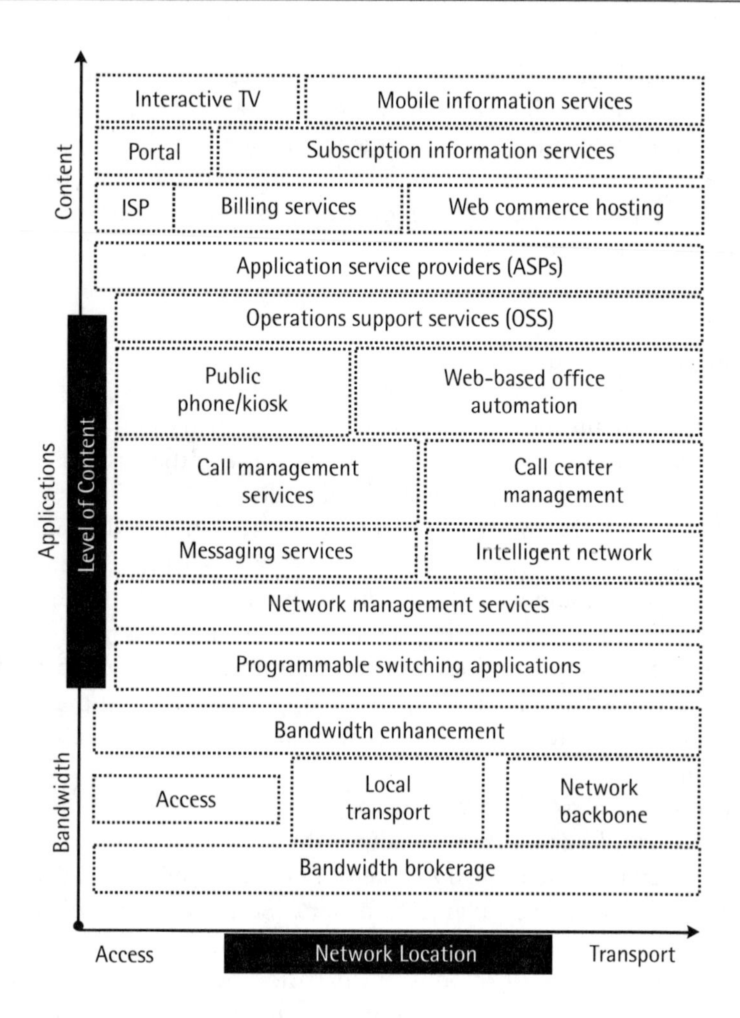

Figure 8.2 New business models.

locally or in a more centralized network location. Companies can innovate in the locations in which they provide service and gain competitive advantage.

The vertical axis in Figure 8.2 moves from bandwidth to applications to content. Bandwidth enhancement refers to service providers that redefine bandwidth and differentiate the service. Examples of bandwidth enhancement that represent active markets are DSL service and cable modem service. These are transitional opportunities, because in both cases, the service provider has modified a waning technology (circuit-switched network or coaxial cable) to meet newer needs. Future networks will feature broadband data-centric facilities from the service provider's hub all the way to the customer.

Today's programmable switching applications include commonly used services such as interactive voice response (IVR), prepaid calling features, and information services (such as sports and lottery information available on the 900 network). Once the network is on a data platform such as IP, a variety of software applications will be available.

Similarly, the market for network management services will grow. Enterprise networks frequently engage outside managers to ensure the reliability of the network. The proliferation of new service providers will create a viable market opportunity for network managers, because most service providers will view that function as necessary but nonstrategic and, therefore, a candidate for outsourcing. AOL, for example, has outsourced most of its network requirements. AOL sold both its ANS and CompuServe networks to MCI WorldCom and currently outsources its traffic on the backbone networks of UUNet, GTE Internetworking, and Sprint [3].

The next tier of network opportunities includes messaging services, call management and other intelligent network functions, call center management, and centralized server-based office automation. Public facilities such as telephones and Internet access facilities combine hardware with applications and can include proprietary content.

The next level of services includes the ASPs and OSS service providers, which capture network or transactional data and provide analysis or information to customers. The architecture of both of these services will probably include centralized software code and some data and other code and data locally available on user PCs. Most software will reside centrally, and for most applications, the applications will distribute the data for security, accessibility, and other reasons.

As the business models move up the chart, they are characterized by more proprietary content, less resident hardware, and more centralization. For example, the models of interactive TV and mobile information services are nearly one-way transmissions of broadband content, customized only by the user's setup requirements and occasional binary transactions.

Figure 8.3 overlays the strategic modeling decisions defined earlier on the new telecommunications landscape, demonstrating that access and transport are most likely to become undifferentiated and use price as

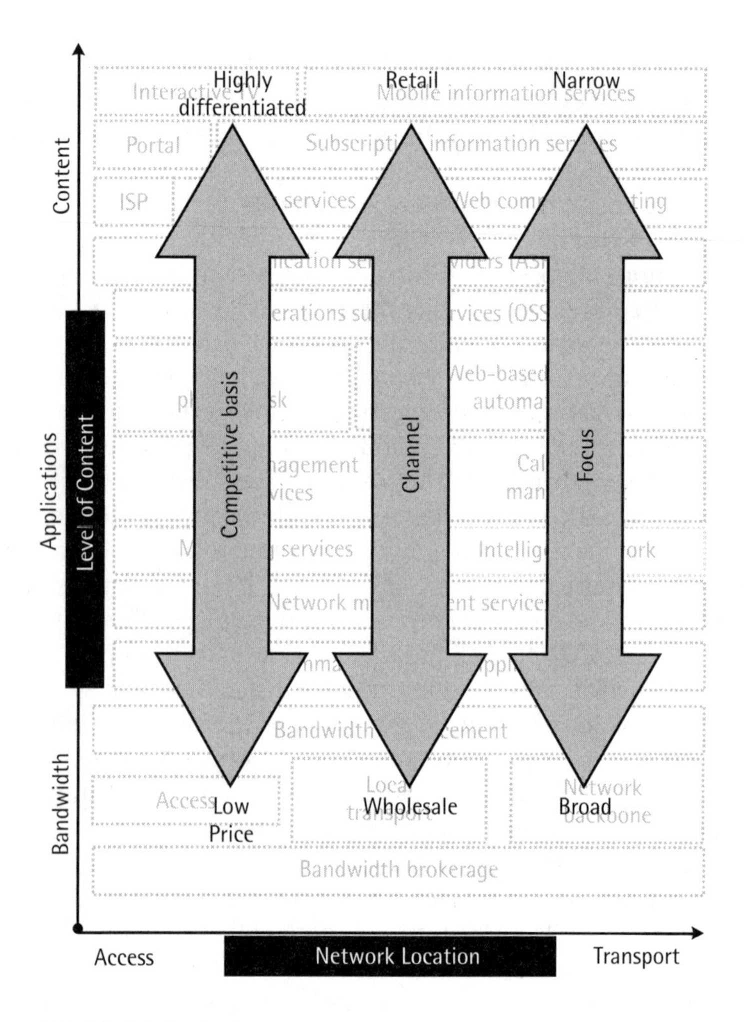

Figure 8.3 Model strategies.

the primary basis for competition. Similarly, economies of scale will be available to providers; the networks will serve a broad customer base through the wholesale distribution channel.

The implication of Figure 8.3 is that service providers that choose to compete by differentiating their services to retail customers will most often select business models that include more content and less focus on bandwidth than in today's market. It also implies that telecommunications service providers that view themselves primarily as bandwidth suppliers will compete most effectively with a low-cost, commodity model. The apparent conclusion is that some of today's network owners, serving or seeking to serve retail markets, have not positioned themselves well to compete against low-cost suppliers in the wholesale market. At the same time, they are poorly positioned to compete against content providers that have wholesale suppliers for bandwidth and differentiated products to offer to customers. It is possible that the largest telecommunications providers support business models that will change the most dramatically when markets are intensely competitive.

It is useful to evaluate the present collection of telecommunications service providers in light of their positioning in the industry and the qualities that make them successful within their chosen business model. Before a service provider considers expanding into a new area, it is appropriate to evaluate its own capabilities to ensure that success is likely. For example, successful telecommunications providers in the bandwidth segment of the industry will undoubtedly excel at reliability, operational excellence, and low cost. In the applications segment, innovation, customer support, and time to market are success factors. In content areas, marketing excellence and user-friendly customer interfaces are critical. Horizontal movement on the chart, from access to transport or transport to access, will require fewer infrastructure changes than movement between segments. A telecommunications service provider considering expansion toward a new business model will need to assess its own capabilities against its likely competitors. Similarly, in the strategic planning process, the astute telecommunications provider will predict the likely new entrants into its existing markets.

If most of today's telecommunications giants uphold a highly defined business model, it is not evident in their mission and vision statements. A vision statement has the potential to encapsulate the business model, describing the value proposition, the target market, if applicable, and the uniqueness of the service provider. Most telecommunications service provider mission statements could be substituted for each other's, demonstrating either that they intend to compete on an undifferentiated basis or that their mission statements do not reflect their business models. An informal survey of U.S. telecommunications provider statements includes phrases such as the following:

- "To be the customer's first choice for communications and information services in every market we serve" (Bell Atlantic);

- "Nothing less than market leadership in the telecommunications industry" (GTE);
- "Dedicated to being the world's best at bringing people together anytime, anywhere" (AT&T);
- "To be the most profitable, single-source provider of communications services to customers around the world" (MCI WorldCom);
- "To be a world-class telecommunications company—the standard by which others are measured" (Sprint).

To be sure, the newer entrants are not much clearer. Global Crossing identifies itself as a "carrier's carrier" but purchased retail telecommunications provider Frontier Communications. Williams Communications, often exhibited for its focus on wholesale markets, sports a twofold strategy. The first targets the market of communications service providers; the second includes "organizations of all sizes." Figure 8.4 demonstrates the likely movement of telecommunications service providers into new markets: Although most of the present industry participants appear to head in the direction of content, telecommunications service providers are expanding along the horizontal axis instead. Unless a strategic shift occurs, incumbent local and long-distance companies are struggling to win the future wholesale market.

8.4 Emerging models

Many new business models have emerged, and some of them are already carving out markets in the telecommunications industry. The following sections will describe several models to demonstrate the value proposition inherent in each and likely success factors for the industry participants.

Bandwidth brokers

Commodities create commodities markets, and bandwidth is apparently no exception. The interexchange portion of the network maintains overcapacity, a trend that will probably continue, even in the face of explosive customer demand. This overcapacity is partly due to the rash of new networks under construction by service providers in the United States and

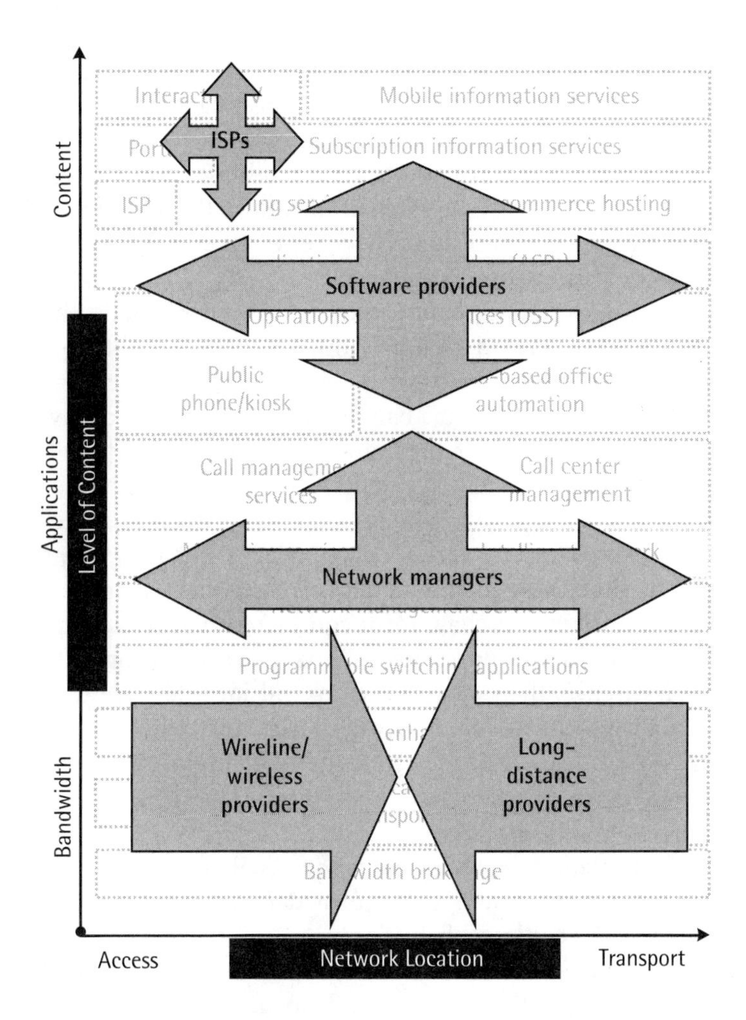

Figure 8.4 Player positioning.

Europe and is magnified by technology advances that multiply the bandwidth of existing networks.

Companies such as RateXchange, Arbinet, and Band-X are establishing a beachhead in this market. Wholesaler Enron has entered the bandwidth trading market as a complement to its other business operations, utilizing a model that was successful in its earlier energy utility ventures. Conventional trading without exchanges was not sufficient to meet the timely delivery of bandwidth demanded by telecommunications service providers. Philips Tarifica estimates that the market for on-line bandwidth and minutes trading will exceed $8 billion in three years, including 10% of all IP traffic [4]. Dow Jones created a bandwidth commodity index for bandwidth to track the prices of network bandwidth. The entrance of commodity brokers into the marketplace demonstrates the widespread belief that unadorned telecommunications service, whether broadband or not, will remain a commodity for some time.

The value proposition for bandwidth brokers is the benefits they offer in reducing the service provider's usage costs and in their ability to provide just-in-time bandwidth. The economies of scale inherent in a commodities market enable brokers to manage bandwidth availability in a way that is simply unavailable to small or medium-sized service providers. Brokers will undoubtedly manage unanticipated or episodic bandwidth demands, even for service providers that maintain their own networks. For decades, facilities-based long-distance carriers have engaged in bandwidth trading and resale to stabilize the demands on their own networks. Brokers provide an efficient and anonymous channel for commerce that long-distance and local service providers will appreciate. As brokers are service providers only in the broadest sense, their success factors will include scale to provide liquidity, a robust information services infrastructure, and sales channels to ensure a constant supply of the bandwidth commodity.

Data CLECs

Data CLECs represent a new business model, facilitated by the introduction of DSL technology. These service providers upgrade existing copper plant to broadband facilities on behalf of competitive access providers such as voice CLECs and ISPs. The best-known data CLECs in the United States are Rhythms NetConnections, Covad, and NorthPoint.

This business model operates primarily within a wholesale market. The data CLECs' customers are service providers, usually ISPs, IXCs, and conventional CLECs reselling the facilities of incumbent carriers. Data CLECs enjoy an uneasy relationship with incumbent local providers. Rarely, data CLECs act as outsourced labor to install DSL facilities on behalf of incumbent providers. More often, incumbents view data CLECs as a reminder of lost opportunities. Nevertheless, SBC has invested in a data CLEC, as has AT&T, which apparently chose to outsource this capability rather than bring it in-house. MCI WorldCom holds a stake in one data CLEC but continues to support others with business as well. Verizon and NorthPoint have merged their DSL businesses, creating a national broadband company. Occasionally, data CLECs develop direct business relationships with large corporations to assist them in establishing remote offices or telecommuting programs. This is still a quasi-wholesale relationship, in that the largest enterprise networks resemble local service providers in their own right.

The value proposition of the data CLEC includes the specialized technical skills they offer and the independence of the wholesale channel. Success factors include reliable customer service and state-of-the-art technology. In contrast to the ISDN debacle, incumbent local providers want to ensure that installation of DSL broadband technology goes smoothly. There is the potential that a singular focus on DSL is dangerous in a technological environment migrating toward packet-switched broadband, but most data CLECs are fiercely entrepreneurial. As the market moves to new technologies, data CLECs will undoubtedly uncover new niche opportunities.

Business-to-business commerce portals

Consumer-directed portals such as Yahoo!, Excite, and AltaVista garner a great deal of attention, but most industry analysts view the potential business-to-business market as huge. The Boston Consulting Group estimated that as much as one-quarter of U.S. business-to-business purchasing will be conducted on-line by 2003, growing by 33% per year to a 2003 transaction value of $2.8 trillion [5]. Less than one-third of that commerce will take place through more traditional electronic data interchange (EDI) over private networks, and their share will shrink. The same study predicted that by 2003, e-commerce penetration will triple from 7% to 24% in North America, from 3% to 11% in western Europe, from 2% to 9% in the Asia-Pacific region, and from 2% to 7% in Latin America. The Yankee Group estimated a more aggressive growth rate of 41% and a less aggressive transaction volume of $541 billion in 2003.

Vendors of ERP applications recognize this business opportunity, and their incentive to increase their user base within existing customers has encouraged them to add portal capabilities to their applications. Corporations that move procurement on-line have the opportunity to control the frequency of purchases and the costs incurred by each transaction. The Yankee Group anticipates that ERP vendors will incorporate four categories of portal services. These include corporate services such as supply management, personal services, consumer-oriented services for employees, trading community management to manage trading relationships with product vendors, and enterprise application extension, on-line purchasing, and selling.

FreeMarkets represents a variation of the business portal model. The company uses reverse auctions for business procurement on the Web. Prescreened suppliers bid on supply contracts, starting from the highest bid and attracting lower bids until a supplier offers the best possible price. Business-to-business auctions are less glamorous than their consumer counterparts on eBay, but they represent a much larger potential market for commodity products, including volume purchases and price-sensitive buyers.

BellSouth's joint venture with business-to-business supplier Commerce One will sell software and supplies to telecommunications service providers and enable their customers to take advantage of consolidated buying and supply auctions. SBC purchased e-commerce software and services vendor Sterling Software. US West partners with vendor Vsource to procure office supplies over the Internet for small and medium-sized business customers.

The value proposition, as for all gateways, is that the portal can attract specifically targeted customer segments. For example, anyone viewing a procurement portal for industrial supplies is most likely receptive to associated business-to-business advertising. Portals also offer superior traffic and tracking mechanisms compared to alternative advertising vehicles. Success factors, besides simple response time, include the portal's ability to maintain relationships with suppliers and customers, and its use of innovative back office technologies.

8.5 Supporting tools

Computer-based modeling tools can assist the telecommunications planner in creating and testing business models and potential scenarios. Because these tools are iterative, managers can alter the variables and assumptions to ensure that the resulting models are robust and feasible.

Balanced scorecard

Many planning tools are available to assist in the business modeling process. One popular modeling technique is the balanced scorecard. The balanced scorecard emerged in the 1990s to address the management concern that strategies often did not translate into actions. The goal of the analysis tool is to develop measures for strategic objectives and communicate strategies throughout the organization. Four perspectives require strategic consideration: financial, customer, internal, and organization learning. The financial perspective monitors the traditional financial indicators such as revenue growth, profitability, and shareholder value. The customer perspective views the organization through the eyes of its customers, measuring customer satisfaction, service levels, churn, and customer retention. The internal perspective reviews conventional efficiency measures such as productivity, cost, and cycle time. The organization learning perspective attends to the future, through innovation, employee skills development, and intellectual capital. One could argue that planners could consider the components of the balanced scorecard tool separately in other domains. The reason that the balanced scorecard analysis resides in the planning domain is that its strength is in balancing the various objectives by linking measures to each other and to financial goals.

Customer segmentation analysis

This planning tool creates efficiency in marketing and sales by targeting customers that value the uniqueness of the offering, driving innovation in areas that will draw buyers, and maximizing achievable revenues and profits. Customer segmentation divides the universe of customers into groups that share common attributes or buying characteristics. The process usually begins with profiling, based on demographic or transactional characteristics. Profiling helps to identify the purchase criteria of customer groups, which makes it possible to develop packages of services, prices, channels, and promotion to persuade customers to buy. Telecommunications service providers are using data warehouses to develop customer profiles for segmentation. The modeling techniques provide direction on issues such as how to retain profitable customers and how to discourage probable churners, or what packages would appeal to demographic segments such as families with young children, college students, or travelers. Customer segmentation applies to the business modeling process, rather than the marketing domain, because the markets targeted—and those excluded—are part of the business model. The segmentation strategy needs broad definition during the planning process.

Scenario planning

Scenario planning intends to provide early warning for business risks and opportunities through a creative planning process. Its advantages are that it offers decision support to help compare strategic alternatives in an environment of uncertainty. The basis of this tool is to develop scenarios or likely views of the future. Scenarios are plausible and focused on the likely drivers of change, including technology, economic trends, political forces, and social change. Once the future scenarios emerge, planners can recommend actions that optimize the anticipated view or minimize identified risks. Forecasting the likeliest scenarios also offers insight into which performance measures validate or challenge the recommended strategies. Scenario planning is an iterative process. As performance data becomes available, its analysis will foster the development of new scenarios, and the process begins again.

Value chain analysis

This analysis examines the value chain, which is the set of activities that takes inputs to the enterprise and delivers service to the customers. Each step in the value chain consumes resources, so it needs to create value at least at the level of resources consumed. The objectives of value chain analysis are to identify the activities and skills that are critical to customer satisfaction and profitability, including the linkages to the value chains of suppliers and customers. Those activities become the focus of effort to ensure that they are as efficient and effective as they can be. Within the business modeling process, value chain analysis is most useful if it helps to identify the opportunities for service differentiation and the business models that make the best use of the value proposition of the service provider.

Customer life cycle value analysis

The objective of customer life cycle value analysis is to calculate the level of financial benefit achieved from a customer through the duration of the customer relationship. Some customers are simply more desirable than others, and the most profitable customers offer a geometric increase in profits over the average customer. The more accurately a service provider can identify these customers, the better it will be able to offer them inducements to be loyal. The lifetime value of a customer includes a host of variables. Acquisition costs can involve commissions, advertising, promotional pricing, and back office account administration. The cost of maintaining the customer includes service provision, customer care, and billing. There are even costs associated with losing the customer, such as decommissioning the service or calculating an exception bill. Without knowledge of the customer life cycle value, investments made on behalf of customers as a whole waste the investments made on unprofitable or undesirable customers. Putting additional resources into maintaining relationships with profitable customers can help to increase desired customer retention and eliminate customers that do not produce profit.

8.6 Role of information technology

There are two major influences from the rise of information technology on the business modeling process for telecommunications service providers. The first is the availability of sophisticated PC-based planning tools, which enable planners to conduct complicated analyses such as those described above with the ability to adjust the elements indefinitely until a useful and accurate model emerges.

The second significant effect of information technology is the impact of the Internet on business models for all companies. This is especially true for telecommunications service providers. For many of the traditional business models in the telecommunications service market, the accessibility of entry to Web-based businesses reduces barriers to entry. The most obvious example is the Yellow Pages directory, which will undoubtedly move entirely to the Web when complete Internet penetration occurs. Hyperlinks, search strategies, and extremely targeted advertising make the on-line version of the Yellow Pages more valuable to both customers and advertisers than its printed predecessor. Like all businesses, the availability of the Internet offers a new and distinctive distribution channel. A unique, not yet imagined Web-based customer service provisioning innovation beyond simple automation of today's practices will be part of some telecommunications service provider's value proposition in the future.

The third noteworthy impact of technology on the business model is the broadening of information technologies into communications technologies and vice versa, fostering convergence. Convergence creates a higher level of competitive intensity, compresses the acceptable time to market for new products, and increases the number of well-funded new competitors. An indirect effect of information technology is to make telecommunications usage more necessary and bandwidth more desirable and to introduce pricing expectations, such as unlimited Internet usage for a price comparable to basic telephone service.

References

[1] Kim, W. C., and R. Mauborgne, "Strategy, Value Innovation, and the Knowledge Economy," *Sloan Management Review*, Vol. 40, No. 3, pp. 41–54.

[2] Henry, B., and D. Rosen, "Formulating the New Business Model," *Telecommunications, International Edition,* Vol. 32, No. 1.

[3] Deng, S., "Why Build, When You Can Outsource?" *America's Network*, Vol. 103, No. 2, p. 78.

[4] McElligott, T., "From Telecom Oddity to Tele-commodity," *Telephony*, Vol. 237, No. 18, p. 94.

[5] Duvall, M., "Report: B2B E-Commerce to Skyrocket," *Inter@ctive Week*, Vol. 6, No. 52.

9

Core Competencies

We do not want to be the water company.

—RBOC executive

The concept of core competence, originally introduced at Harvard [1] in 1990, has become entrenched in the business lexicon. Like so many other strategy innovations, explorations of the concept by companies or industry observers sometimes suffer from a lack of depth or a disconnection between analysis and implementation. Core competencies are evident either in skilled operational processes conducted by the business or in knowledge held within the enterprise.

To qualify as a core competence, the capability should have three characteristics:

- Provide a significant value to customers;
- Be competitively unique and relatively difficult to imitate;
- Offer a basis for entry into new and expanded markets.

Based on these characteristics, it is apparent that some telecommunications service providers apparently presume that their skills qualify as core competencies when they do not. Simply providing excellent service is not a core competence unless it offers unique value, differentiates from competitive offerings, and provides a gateway to new markets. Network reliability in the facilities-based interexchange market is not a core competence for established circuit-switched providers, although it could open new markets for facilities-based IP service providers. Resellers should focus upon core competencies outside of technology development unless they have access to proprietary development processes. On the other hand, telecommunications service providers seeking to enter new markets—and most of the largest providers consider this a priority—will facilitate their market entry if their reputation for demonstrated core competencies precedes their market entry.

Many successful organizations do not demonstrate first-rate core competencies, but those that do so command brand premiums and customer loyalty. Core competencies are most often associated with business processes rather than specific services or patents on technologies. Intel's business processes supporting its 18-month cycle between each new chip are its true core competence, not the patent on any specific chip. Patents are temporary, but business processes are sustainable.

The business processes demonstrating core competencies must be central to the operations of the enterprise. Superior office design is probably not a worthy core competence for telecommunications service providers; superior Web site design most often is. MCI's core competence in its first two decades was its excellent applications of information technology to its marketing and operations processes. While MCI WorldCom continues to nurture its information technology capability, in the last five years other service providers have closed the gap. Similarly, WorldCom's announcement that it would review the size of MCI's marketing budget when it announced that it was proud to acquire MCI's valuable brand surprised industry observers. Since then, WorldCom has apparently become more comfortable with MCI's large advertising budget and the brand recognition it continues to produce. BellSouth's core competence is its customer care, as evidenced by its J. D. Power and Associates recognition year after year. Whether customer care will be central to its ultimate business model is difficult to predict, but in the meantime, BellSouth is in position to serve the retail market more effectively than others can with less noteworthy customer care.

9.1 Identifying real strengths

Telecommunications service providers that seek to develop core competencies will need to evaluate their present skills and positioning in an objective assessment. Superior performance compared to competitors requires independent confirmation. In-house customer satisfaction surveys are not always sufficient proof of superiority. The results of these surveys are limited to the perspectives of one's own customers, who may not have experience with other service providers. Furthermore, high ratings from customers offer no comparison with ratings of other providers, which might well be even higher. There are several approaches to evaluating existing core competencies, depending on the investment a service provider is willing to make. The first, and least expensive, option is to seek out trade press that analyzes the company's operations as long as the source of the article is not a recent press release. Ideally, service provider in-house analysts and management monitor all telecommunications trade publications, which occasionally commission their own research to rank various suppliers of telecommunications services. For example, computer publications routinely evaluate various ISPs on price, response time, and other variables that do not constitute core competencies. They also conduct evaluations of areas that can act as core competencies, such as calling technical support with a user problem or judging the on-line content of portals or other Web sites. These studies serve to demonstrate the market position of various service providers and are often searchable in the on-line archives of publications. While similar studies of other telecommunications services are not as plentiful, abundant competition in local markets will undoubtedly increase their frequency. A second approach is to commission market research to ascertain one's own core competencies and those of one's competitors. As this is primary research, this alternative can be costly. The expense can be contained if internal company analysis determines which variables are worthy of concentrated analysis—if the internal analysis can be adequately objective.

Service providers also need to appraise whether any competence-based lead is sustainable against competitors seeking the same capabilities. If, for example, the core competence is deeply dependent upon information technologies that any competitor can package, the competence is vulnerable. Even knowledge-based competencies resident in the organization's staff are susceptible to competitive recruiting. Still, if a competence is uncommon in service providers or even in other industries, and if it adds unique value to the customer experience, it will create sustainable competitive advantage.

Sometimes a telecommunications service provider possesses a unique competence and overlooks the most lucrative opportunity to gain its advantage. When the incumbent LECs priced their electronic directory listings beyond the customer's willingness to pay, they lost the opportunity to create a new service and created a market of competitors in their own directory services niche. As with some other services, they instead chose to keep the retail directory listings business for themselves, creating a new channel conflict with business customers and resellers.

Strategy consultancy McKinsey and Company identifies three classes of capability: privileged assets, growth-enabling skills, and special relationships [2]. Privileged assets confer competitive advantage because new competitors find it hard to replicate them. Examples are the fiber networks owned by AT&T, WorldCom, Sprint, Level 3, IXC (now Broadwing), and Qwest; Microsoft's access to software development for its network offerings; and the RBOCs' powerful brands within their traditional territories. Growth-enabling skills, generally financial capabilities, are those that reach beyond the specific industry to create and sustain growth. WorldCom's ability to integrate acquisitions, the creative financing abilities of new entrants such as data CLECs, and Bell Atlantic's deal-structuring prowess are examples of growth-enabling skills. Special relationships encompass ties with existing customers and suppliers. For telecommunications service providers emerging from the regulatory umbrella, relationships with regulators will ease their transition to full competition. Bell Atlantic's authorization to offer long-distance service in New York State was the first of all the RBOCs. This was partly due to the competitive intensity within New York, partly due to the service provider's compliance with conditions, but also partly due to Bell Atlantic's ability to work closely and negotiate effectively with state and federal regulators. The importance of the ability to work with regulators will wane as the industry becomes competitive, but some of the skills held by Bell Atlantic are transferable to other negotiation situations.

Core competencies require maintenance, and they will disappear through attrition if the enterprise does not apply sufficient resources to their continued development. Part of the maintenance program is to monitor competitors' progress in their own development of competencies, and part is to ensure that management fails to recognize the loss of staff or other critical resources before the loss is unrecoverable. The most important aspect of maintenance is to recognize which business processes represent actual or potential core competencies, so that management can direct resources and attention to them. Part of this effort is to reduce any excess of resources applied to business processes when they do not represent potential differentiators.

9.2 Core competencies by business model

Different core competencies work best for each business model, because the source of customer value varies for each one. Table 9.1 depicts a sample of the business models likely for the telecommunications services industry and some suggested areas in which core competencies could differentiate a service provider. Note that certain competencies that are not obvious differentiators for the general market could be central to a business model targeting a submarket of a larger model. Whether a service provider has industry knowledge generally does not matter—for example, if the customer is a dial-around reseller of commodity bandwidth from a wholesale provider. On the other hand, it can be a significant opportunity in certain submarkets, such as airline reservation networks. Similarly, the fact that an ISP owns a worldwide network would not matter to most customers. It would matter considerably, though, to the multinational corporation that seeks an ISP to host its intranet with a standard log on and a low operating cost. Thus, the core competencies that provide the most advantage relate directly to the specific value proposition offered by the telecommunications service provider to its well-defined market segments. In all cases, nevertheless, certain competencies that are necessary but not sufficient will not add unique value in either the most general markets or the most targeted customer segments. Some telecommunications providers have made the strategic mistake of assuming that competencies that are necessary but not sufficient act to make their services unique. They appear not to notice that all successful providers sustain the same competence in the same competitive area. Some service providers base their strategies on excellence in some capability, which most often is valuable to a subsegment of the market. To make this strategy work, the service provider must demonstrate a perceptible superiority in the capability. In addition, superiority in the competence must be important enough to customers to pay a premium, and there must be enough customers willing to pay the premium to cover its total cost.

Although the list of core competencies in Table 9.1 is not comprehensive, it demonstrates a few characteristics of emerging competition in the telecommunications market. For example, some competencies about which service providers currently boast are not points of differentiation but rather baseline requirements for all providers. The most visible example is the reliability of the network, a requirement in most telecommunications markets. Very good reliability is simply not a differentiator in today's market, and it is unlikely that competitors will emerge with inferior reliability and survive against the existing providers. Perhaps small niche markets will develop for superlative reliability, as did the parallel processing,

TABLE 9.1
Business Models and Associated Competencies

BUSINESS MODEL	NECESSARY BUT INSUFFICIENT CAPABILITIES	POTENTIAL DIFFERENTIATING CORE COMPETENCIES
Access and transport	Service reliability Cost management	Large network footprint Market segmentation and targeting Just-in-time provisioning
Bandwidth enhancement	State-of-the-art technology	Innovative packaging
Programmable switching	Network reliability	Technology innovation User friendliness
Network management	Network reliability	Information reporting technology
Intelligent network functions	Rapid time to market Network reliability	Market segmentation and targeting
ASP and OSS providers	Availability and response time Security Cost management	Technology innovation
ISPs, portals, and interactive content providers	Service reliability State-of-the-art technology	Customer service On-line content Strategic alliances Innovative pricing

extremely reliable computers in the information technology market, used primarily for automated teller networks and other high-priority applications. For most providers, reliable service is simply the cost of market entry. Other common opportunities are observable in the list. Most markets offer differentiation opportunities through innovative technology, which means that most markets will not exclude smaller providers if they can develop a technology or a feasible segmentation strategy.

9.3 Identifying market opportunities

The analysis and development of core competencies can work in an iterative manner with the business modeling process. The normal course of events for a new business is that an entrepreneur conceives a value proposition and a business model, then sets about to develop the core competencies required for differentiating the service. Sometimes, though, a going concern recognizes a core competence that can act as a differentiator, then seeks business models that can make the best use of the capability. Amazon.com began as a bookseller, in what is largely a commodity market, as demonstrated by Amazon's initial emphasis on discounted prices. Among the core competencies held by Amazon.com is its purchasing infrastructure, including its shopping cart features, Associates agents program, and One-Click® technology, a customer's ability to store personal information for recurring purchases. The courts have protected the company's proprietary rights to this user interface. Amazon.com leveraged the core competence in its infrastructure by adding additional commodity shopping areas such as electronics, music and videos, and other well-known consumer items. It is worth noting that the extended markets Amazon chose to pursue are also largely commodity consumer items. To identify new opportunities, telecommunications service providers can assess their competitive advantages within the strategic planning process, then develop new service lines for which that advantage can serve as a core competence.

Williams Companies, one of the world's biggest pipeline companies, discovered an expansion opportunity because its costs drop when it builds more facilities, whether they are for natural gas transport or fiber optic communications. Williams translated its findings into a business model that targets large business customers, although it has the resources to serve universal markets and the discipline to limit its distribution to the wholesale channel. Knight-Ridder tried to extend its publishing capabilities into the on-line business services market. In 1983, it formed the Business Information Services division and consolidated operations in commodities news, stock quotations, and a very early on-line information service. The division grew through investment and acquisitions, but Knight-Ridder eventually sold its business information division in 1997. Its more recent on-line focus is the RealCities network, an Internet portal directed at consumers. This new focus demonstrates that Knight-Ridder has determined that its best opportunity is to leverage its core competencies in on-line services at its primary print market, the consumer. Not surprisingly, other consumer-directed publishers and content providers also created portals, such as NBC with snap.com, and ABC with go.com.

Brand extension, sometimes called line extension, enables a service provider to leverage the reputation of an existing service to sell a related new service. Most industries use brand extension as a source of growth. It is often successful not only because of a company's brand but through the business processes that made its original product a success. Telecommunications service providers are practicing line extension when they offer Internet access, new enhanced services, or new services under their brand name. Line extensions in these areas are more likely to be successful than new products and services that do not use the same business processes as the original branded services. Examples of the latter are Ameritech's security services, AirTouch's vehicle tracking services, divested in 1995, or the tepid performance of most retail centers of local service providers. Historically, telecommunications service providers have been quite successful at company brand image and much less successful at creating branded service lines or branded services that offer extension opportunities. To extend a brand successfully through line extension, several elements need to be present. First, the business process that creates the real brand differentiation in the original service needs to be present in the extended service and needs to create a similar, perceptible difference. Second, the new service needs to link sufficiently to the old service that customers are willing to assume that the new service will be as trustworthy as the brand it extends. AT&T made this error when it acquired the personal computer division of NCR. AT&T and IBM recognized decades ago that computers and communications would converge, and both were generally unsuccessful at horizontal integration and brand extension. IBM's blunders were in the areas of telephone equipment and networks, including an ISP, businesses that IBM has exited by now to concentrate on its e-business initiatives, a successful brand extension. AT&T offered the NCR personal computers under the AT&T brand, including computer-telephony integration early in its development. The AT&T brand was not strong enough to compete successfully in the markets the company targeted, and AT&T quietly spun off the computer division at the same time it divested its equipment manufacturer Lucent.

9.4 Outsourcing

Outsourcing occurs when a business chooses to purchase a service rather than provide it in-house. Telecommunications service providers are often on the selling side of outsourced services, in the fields of network management services, ASPs, and hosted Web services. In the future, service providers will find that investment opportunities that directly increase markets and revenues are so abundant that they will choose to outsource functions instead of operating them in-house. Outsourcing represents a cultural divide between incumbents and entrepreneurs. Entrepreneurs often have neither the capital nor the customer base to justify establishing many in-house business processes. They use service bureaus to handle their payroll processing, "lease" employees from companies to avoid the cost of a human resources management structure, and resell facilities until their customer base justifies buildout. As described in Chapter 6, there are advantages to building and advantages to resale. Similar advantages, such as flexibility and the ability to match costs to revenues, apply to outsourcing certain business functions. In the context of core competencies, though, outsourcing takes a more significant role. All enterprises most certainly should retain in-house any business function that serves as a core competence. Core competencies need management and control, and telecommunications service providers cannot run the risk that their competitors derive equal access to core competencies. Similar reasoning, though, concludes that any function that is not a core competence is a candidate for outsourcing. First, outsourcing frees up both capital and management attention that should be directed to the core competencies of the enterprise. More importantly, it is worth noting that services purchased through outsourcing are undoubtedly the core competence of the provider of the outsourced service. A billing service provider will most likely have a state-of-the-art system and many sophisticated information analysis features. These features are probably not cost-effective to build and maintain in-house. They are, of course, equally available to competitors that outsource the function as well. This is not a detriment, as any function that a service provider can outsource is, by definition, not a competitive differentiator for the provider's chosen business model.

Culturally, incumbent service providers have a larger hurdle than entrepreneurs do. First, an incumbent provider generally has access to the capital required for providing its entire supporting infrastructure, so it becomes tempting to do so. Second, the customer base of incumbent providers is more stable than that of a new entrant, so flexibility to enter and exit markets is less of a concern than low cost, which inclines the provider toward in-house services. Third, regulated incumbent providers operated within a vertically integrated value chain, so the culture is predisposed to serving its needs in-house. This generally means that incumbents already have systems and functions in operation and that they are highly customized, serving as a barrier to exit. Furthermore, when only one vertically integrated supplier serves each market, few incentives exist for the market entry of companies that can provide outside services. The network operations center (NOC) is an example of a business model that has arisen to monitor the status of CLEC networks on an outsourced basis. The NOC monitors network elements such as transmission, switching, and signaling, and alerts technicians to problems or potential service problems for resolution. Facilities-based carriers and resellers can benefit from network management services. Some telecommunications service providers provide this function in-house. Time Warner Telecom established its own center in 1994 simply because outside services were not available [3]. Now, that function would be a candidate for outsourcing.

According to one study, 60% of service providers currently outsource services, and they expect that percentage to increase to 74% by 2001 [4]. The companies surveyed used outside parties to offer critical business elements such as customer care and billing (32%), network planning and construction (28%), OSSs (25%), network integration (20%), and service creation and customization (19%). Non-U.S. companies above $25 million in revenues outsource at a higher rate than their U.S. counterparts do, at 64% to the present percentage of 55% for U.S. companies of the same size. Note that companies are quite willing to outsource critical business elements, and their plans to increase the level of outsourcing demonstrate their satisfaction with the results. This confirms that telecommunications service providers believe that they can entrust vital processes to outside suppliers if excellent performance is essential but differentiation is nonstrategic.

Many telecommunications service providers already use outsourcing in both expected and unexpected areas. Several major ISPs have chosen to outsource their network management function to wholesalers to focus on other priorities. One might assume that ISPs would consider managing the infrastructure of their networks to require a core competence, but the best-performing ISPs in terms of cost per subscriber—AOL, and now-merged Earthlink and Mindspring—are outsourcers [5]. The "virtual ISP" has emerged as an entry strategy for any company desiring to become an ISP. PSINet acts as a wholesaler and handles all the networking and backbone requirements. For the new ISP, using a wholesaler offers features that would be impossible without outsourcing, such as nationwide access points. Wireless operators have begun to sell their towers, redirect the capital, and outsource the management of their cell sites, including design, deployment, and network management [6].

Sprint upgraded its business-class e-mail service to an open Internet standard through an outsourcing arrangement with Critical Path, a two-year-old provider specializing in business e-mail. Sprint sells the service under its own brand, and the combination of lower cost and excellent service made the opportunity attractive. Other local service providers have outsourced or are considering outsourcing e-mail, which is often a high-cost, high-maintenance requirement. Outsourcing e-mail and other enhancements to what would otherwise be commodity service is one way that telecommunications service providers can differentiate their offerings while keeping costs as low as possible. E-mail wholesaling also creates a new wholesale business model for new entrants to the industry that do not have the capital and other resources to compete with the existing giants.

BellSouth outsources 10–20% of its outside plant engineering work, and the percentage could increase as BellSouth refines its outsourcing strategy [7]. GTE has worked with its own unions and outside contractors to strike a balance between in-house and contracted outside plant construction services. Part of its outsourcing strategy included redefining work processes, modifying performance measurements, and decentralization of some decision-making.

AT&T Wireless Services extended a billing and customer care services contract until 2004, representing about $1 billion in revenues. Customer care outsourcing represents one of the functions that are very cost-effective to outsource for small-to-medium-sized enterprises—and sometimes for the largest companies as well. Because customer care and billing require 24-hour availability, most companies with in-house customer care centers experience very slow periods. These periods require staff, management, and all fixed business costs. An outsourced customer care operation transfers these costs to a service bureau that can optimize its resources to meet the needs of many customers.

9.5 Seeking and targeting competitor weaknesses

The reverse of leveraging core competencies is to identify competitor weaknesses and create markets that exclude competitive entry. This is most effective in oligopoly markets, in which a service provider can easily tally competitors, and their weaknesses are more apparent than in diverse markets in perfect competition. A telecommunications service provider can create a competence by identifying a common weakness among all competitors and then nurturing a capability that competitors are unable or unwilling to match. One potential example is in the area of local wholesale services. The present list of facilities-based local service providers includes wireline and wireless, cable and copper, and large and small providers, but nearly all providers are also retailers of their own facilities. This leaves resellers, a growing segment, with a market need to purchase facilities to resell from a supplier that is not also a competitor in the retail market. This channel conflict creates a distribution opportunity to develop a core competence that today's competitors are apparently unwilling to match. Facilities-based carriers with retail customers have the benefits of vertical integration and the profits associated with retail markets. Few have considered splitting the wholesale and retail portions of their businesses, and of those, none has gone through with a sustained organizational restructuring. A facilities-based local provider with a commitment to wholesale can gain core competencies in serving the wholesale market, which is a peripheral business to most facilities-based providers. Superior service coupled with no channel conflict can produce a profitable business model for a service provider that is willing to make the commitment.

AOL used its commanding lead in families on-line to create proprietary buddy lists and instant messaging services for consumers. Even if its competitors could create software with the same capabilities, as MSN did, AOL's huge customer base made the AOL version of the service worth much more. The weakness AOL exploited is simply the small base of customers that competitors support. Competitors cannot readily acquire the uniqueness of the AOL service because they simply cannot acquire the customers. The concept of "universal service," the cornerstone of telecommunications pricing for a century, was originally predicated on the concept that communication is most valuable when more customers are on the network. AOL reaps the benefits of its large user base by creating innovative messaging services and then educating users to need them. Other examples of telecommunications service providers exploiting the weaknesses of their competitors include successes described in Chapter 3. MCI leveraged its information technology skills in the Friends and Family program and changed AT&T's large customer base from a competitive asset to a burden. AT&T was able to revolutionize wireless pricing because its competitors owned neither a nationwide footprint nor a backbone long-distance network. A lack of land-based infrastructure prevented competitors from gaining the economies that enabled AT&T to eliminate long-distance and roaming charges. Therefore, telecommunications service providers can differentiate their services either by possessing unique competencies that are unavailable to competitors or by highlighting weaknesses that are common to competitive service providers.

9.6 Supporting tools

Because the area of critical self-assessment is so difficult, tools can help to eliminate some of its subjectivity. The qualitative tools are targeted and objective, and the quantitative tools can offer more rigor than a purely judgmental analysis.

Core competencies analysis

A straightforward core competencies analysis explores a series of questions about the business. The questions include the following:

- What are the specific processes, skills, knowledge, or other assets that create a unique benefit to customers?
- How can the organization use these competencies to retain customers or draw customers from competitors?
- How can the organization use these competencies to create new markets?
- What actions will make it difficult for competitors to acquire or replicate the same competencies?

- What investments are necessary to ensure that the competencies remain unique and valuable?

Core competencies analysis requires the organization to meet several conditions to create an environment for successful implementation. Like many strategic processes, senior management must be knowledgeable and supportive of both the analysis phase and the conclusions it draws. Executives must be willing to discard some of their intuitive beliefs about the success factors of the industry and the strengths of the enterprise. They must entrust divisional and operational management with tools and a directive to develop needed competencies, imposing limits on capabilities that are not deemed core competencies. This creates cultural hurdles throughout the organization. Some organizations are understandably reluctant to harvest a superior-performing business process that the organization no longer regards as a differentiator. One solution for this situation is to divest the excellent division and purchase its services on an outsourced basis. This enables the newly freed division to concentrate its investment and management on the function that has become its own core competence. It also allows the divesting organization to redirect the capital that would have funded the spin-off and apply it toward real market differentiators.

Consulting firm Booz-Allen & Hamilton suggests a three-phase process to evaluate core competencies [8]. Its first phase is to understand and identify the core competencies driving the telecommunications industry. In the second phase, the service provider selects the competencies needed and positions them against those of competitors. The third phase involves developing the critical core competencies using management processes, information systems, corporate structures, human resources, and performance systems.

EVA

Economic value added (EVA) analysis can be a valuable tool to determine which business functions can serve as core competencies. Similarly, it can help to identify functions that the organization can outsource successfully. EVA is a performance measure similar to traditional, less comprehensive metrics such as net present value (NPV). While EVA is most often applied for a total company financial analysis, it is most useful to evaluate the performance of divisions or business functions. EVA's unique distinction is that it considers the opportunity cost of capital in its analysis, not simply paper profits. The importance of EVA analysis in the strategic assessment of core competencies is that it highlights those business processes that add genuine economic value to the enterprise. The calculation of EVA takes the net operating profit and subtracts an appropriate charge for the opportunity cost of capital invested. By including the cost of capital, a real and sometimes large cost in the telecommunications industry, the service provider can reveal whether a business function adds or subtracts from overall profit. Management is sometimes surprised to discover that the addition of capital costs changes the profile of a particular division or function from an acknowledged success into an area needing attention. The EVA analysis can also validate new candidates for outsourcing, because outsourcing almost never requires capital investment, so those costs are not applicable. EVA can focus management on its most profitable opportunities and change the management culture to oversee assets as well as revenues and costs.

9.7 Role of information technology

Information technology plays a less prominent, but still important, role in core competence strategies than in most other strategic planning functions. First, as with other techniques, information technologies enable analyses, especially iterative analyses, that were unthinkable when computers were not readily available and software was less sophisticated. As in other planning applications, the value of the information technology is in the breadth of its consequences rather than the amount of activity in its application. For example, EVA analysis would be impossible without the ability to run complex applications, even at the most general company level. Its value to core competencies analysis requires a more detailed evaluation of specific business processes, lines of business, or potential new markets. Customized applications are essential to its ongoing analysis, measurement, and decision support.

Second, information and communications technologies promote outsourcing. Organizations can redirect customer service lines to a service bureau or toggle back and forth from in-house staff to outside support during nonbusiness hours or peak service periods. Similarly, real-time transactions can take place at any location, so management can make the outsourcing decision primarily on the costs of each alternative. Last, technology-based performance measurements enable management to oversee outsourced functions with confidence. Managing outside employees with the support of an extranet is not very different from managing over an intranet. With the anticipated growth of telecommunications usage, coupled with the increased need for bandwidth and the expansion opportunities in new wireless or global markets, no telecommunications provider should hold any function in-house that does not offer a competitive differentiation.

Third, the management and operations of information technology represents an obvious candidate for outsourcing on its own. This is especially valid for incumbent providers, whose original computer applications, such as billing, still contain legacy elements from decades ago, when pricing, services, and geographical scope were decidedly different. Legacy information systems are often erroneously viewed as core competencies when management fails to differentiate between a large amount of sunk investment and the associated uniqueness of a platform and code with a perception of value. Furthermore, information technologies supporting telecommunications service providers are among the fastest growing segments of the market. Provisioning and OSSs are available from dozens of developers with specialized industry knowledge, and for which information technology is a core competence. Moreover, telecommunications service providers can outsource information technologies at many levels, from in-house purchases of systems developed outside to the use of ASPs for unaffordable applications, to the use of a service bureau for a truly turnkey operation. Few telecommunications service providers will continue to maintain an in-house staff of programmers and analysts at anything near the scale of the days of mainframe computing. Outsourcing information technologies is often the most cost-effective choice, but it also provides needed flexibility to telecommunications service providers that will need to move faster than they ever have in the past.

References

[1] Prahalad, C. K., and G. Hamel, "The Core Competence of the Corporation," *Harvard Business Review,* Vol. 68, No. 3, pp. 79–91.

[2] Baghai, M. A., S. C. Coley, and D. White, "Turning Capabilities Into Advantages," *The McKinsey Quarterly*, 1999, No. 1, pp. 100–109.

[3] Gary, K., "Build or Buy," *X-Change*, Vol. 3, No. 7.

[4] Biagi, S., "Telecom Tarot," *Telephony,* Vol. 237, No. 23, pp. 32–36.

[5] Deng, S., "Carriers Go Further With a Little Outsourced Help," *Telephony,* Vol. 236, No. 4, p. 64.

[6] Gohring, N., "Outsourcing Gains Ground," *Telephony*, Vol. 237, No. 2.

[7] Lindstrom, A., "Price, Performance, & Productivity," *America's Network*, Vol. 101, No. 4, p. 66.

[8] Booz-Allen & Hamilton, "The Telecom Future: Core Competencies for the 21st Century," *Insights*, Vol. 3, Iss. 4.

10

Organizational Strategies

Take my assets—but leave me my organization and in five years I'll have it all back.

—Alfred P. Sloan

Incumbent telecommunications service providers have made great progress in transforming their organizations to succeed in a competitive environment. Still, more is necessary. The vast size of the incumbents, coupled with cultural factors, has served as an obstacle as well as a facilitator of competitive entry. When incumbents were monopolies, they kept their service and prices at levels that satisfied customers and regulators, and maintained a sporting interest in the performance of their peers. In competitive markets, customer satisfaction is often insufficient, and maintaining superiority, not simply parity, with competitors is critical.

Mercer Management Consulting conducted a productivity comparison of 72 telecommunications operators in 59 countries [1]. Its major conclusions included the fact that operating efficiency and service quality vary dramatically between service providers. In fact, the consulting firm estimated that $78 billion per year in savings in addition to higher service levels were achievable if best practice levels were universal. Recognizing these opportunities, telecommunications service providers have made considerable investments in organizational strategies, and they will continue to do so as competition intensifies.

10.1 Company culture

The culture of a corporation comprises its core values. Corporate culture affects how employees approach their jobs, how they approach risk, and how they operate in teams. Culture affects the ways in which managers reward excellence and repair organizational problems and whether management empowers employees to make decisions or consume corporate resources when customer service is at stake. No culture is inherently better than another is. Some cultures match monopolies well but not other industry structures, and these cultures will need to evolve when deregulation introduces competition. Within an industry, some values work for one company but not its competitors. For example, empowering employees became popular in the last decade. Decentralizing decisions to frontline employees is widely considered a good idea but not if it changes hiring practices to require highly skilled, expensive clerks and service representatives instead of lower-paid personnel that get the job done well. When one customer receives special attention from a call center representative, it is wonderful for that individual customer but a potential service problem for the five other customers left on hold while the on-line customer enjoys the service experience.

Certainly, the issue of corporate culture is very significant when moving from a regulated to an unregulated market. Bell Atlantic approached the problem head-on with its development program "The Bell Atlantic Way." Managers learned new ways to make purchase decisions, coach their subordinates, and even coach their peers and leaders. The learning process involved discussions about the looming competitive business situation, but it also covered a set of principles that would enhance most organizations. The cultural transformation process included a set of core values and language to describe the new management processes.

While the movement to competition is among the most visible requirements for cultural transformation, other significant changes in the industry are creating cultural challenges. Industry consolidation is joining large companies with small ones, and veterans with mavericks. Mergers like the ones between Bell Atlantic and NYNEX or MCI and WorldCom are more likely to succeed than would those where stranger bedfellows are united. Cultural issues would undoubtedly have hindered, if not derailed, the attempted mergers between MCI and British Telecom or between Bell Atlantic and TCI. Mergers between relative equals are more vulnerable to cultural issues, and the dissolutions of the mergers between Telenor and Telia and the one between KPN and Telefónica probably portended cultural challenges when the mergers completed. Traditional telecommunications service provider US West overcame some cultural hurdles when it acquired entrepreneurial cable provider Continental Cablevision. The two companies had public disagreements about strategic priorities and the location of the merged headquarters. US West invested significantly in integrating the two companies before it eventually sold the MediaOne division to AT&T.

Ironically, the same concerns that surrounded the potential merger between British Telecom and MCI appeared at the announcement of the MCI merger with Sprint. This time, though, observers characterized MCI as the straitlaced traditional partner in conflict with free-spirited Sprint.

ISP Verio used acquisition as its entry strategy and then initiated the integration of 16 regional ISPs located all over the United States. Recognizing the inherent difficulties in integrating locally developed back-office systems, the company uses purchased billing and OSS software. In-house systems competing for a merged infrastructure create a winner and a loser; purchased software is neutral. Furthermore, developers specifically design off-the-shelf software to meet the diverse needs of prospects and customers, meaning that processes in acquired companies can be phased in at a leisurely pace, when necessary. Software can thus ease the cultural transition to the merged entity.

Culture represents a primary challenge to telecommunications service providers contemplating globalization. Companies often report meetings in which the participants from various countries walk away with different perceptions of the meeting's outcome and action requirements. Work environments and employee reactions to management pronouncements are only two of the many sources of cultural confusion that can pervade a global business organization. For the telecommunications giants, resolving cultural conflicts is a necessary hurdle to meeting their growth objectives.

One more source of cultural conflict arises from distribution channels. Historically, the largest telecommunications providers vertically integrated their businesses, including the manufacture of network components, through network service provision, through retail marketing. Because they controlled the distribution chain from end to end, the culture was also quite homogeneous. Mandated breakups and good business sense have decimated the distribution chain. Large, traditional suppliers will need to gain the skills to work well with entrepreneurial customers; large service providers will need to expand their sourcing to include innovative but small suppliers. Companies that limit their distribution networks to partners in similar cultures are creating self-imposed obstacles compared to those that find ways to work in disparate cultural settings.

10.2 Knowledge management

The rise of the knowledge worker has prompted organizations to consider how to capture the knowledge resident in the organization. In an industrial age, companies replace retiring machines by other machines that perform the exact same tasks. Knowledge workers are irreplaceable when their knowledge derives from their specific experiences. Turnover, attrition, and the sheer size of service providers all erode the potential knowledge foundation of the employee base. The largest telecommunications service providers could have hundreds of thousands of employees, so their ability to communicate seamlessly is required, or they are at a competitive disadvantage with the small companies with a single location and constant personal communications. Companies dependent on the knowledge and experience of workers are willing to make investments to capture as much knowledge as possible from their most experienced workers. In fact, many knowledge-based organizations use knowledge management systems routinely simply to disseminate the knowledge from active employees to the entire organization. Knowledge management is also useful in companies that require collaboration among employees in different locations or disparate time zones, a problem that will increase in importance as the global industry becomes more consolidated.

Once technology and organizational design collided to create knowledge management, many telecommunications service providers created a senior-level management position to ensure that the knowledge management programs operate effectively. British Telecom, AT&T, and Deutsche Telekom are among the companies that have created chief knowledge officer positions. Whether these positions hold any decision power is a private company matter, but the fact that the titles exist demonstrates that these service providers believe that intellectual assets have value.

Knowledge management systems must include several components to be most effective. Information technology support needs to be adequate in both scale and communications response time, and the database should include user-friendly search capabilities. Nontechnical users who are unable to capture the data they need quickly will abandon the system and waste its considerable investment. The tools in the search engine need to pinpoint the proper information when requested. Processes need to support the facilitation of information retrieval and must be in place to assist in the creation of new information as it becomes available and the deletion of outdated or inaccurate data. A knowledge management system, like most management information systems, is a living repository. Decisions about content are among the most difficult and subjective. Too much data will result in extraneous retrieval and harm the productivity of employees; too little data will disappoint employees looking for answers and discourage them from using the system frequently. It is useful to maintain system performance metrics, which can help to determine the criteria for new data to enter the system. The type of data to be available must pass tests defined in the design phase. The data should be limited to information that will increase the performance of employees or improve the customer's experience. Management and cultural issues also affect the success of knowledge management efforts. Individuals holding unique knowledge are sometimes reluctant to distribute it freely; effective incentives and supportive core values can encourage the most expert employees to share their knowledge.

While the benefits of knowledge management systems appear to be too theoretical to measure, new metrics are available for determining the returns on knowledge management systems. Quantification of benefits is most visible in customer service organizations such as sales and customer support. For example, a customer service center could use a knowledge management system to help service representatives to identify the source of problems by listing troubleshooting measures that were successful in the past. Knowledge management metrics could show that more problems are resolved with a single call in customer service centers that use on-line staff support than those that are not similarly equipped. Telecommunications service providers have used knowledge management systems to increase their sales productivity. Sales representatives tend to specialize in those services that they have sold successfully in the past. Knowledge management systems can help to increase sales by providing information about services with which the sales representative is less familiar. Furthermore, the knowledge management system can assist the sales representative in recognizing sales opportunities for services outside of the norm.

One component of knowledge management is the data warehouse, or data mining. Most large telecommunications service providers support a data warehouse, and several, such as MCI WorldCom, have made a considerable investment in this technology with the objective of achieving competitive advantage. By 1996, 40% of service providers had already installed a data warehouse, while another 40% of the remaining carriers expected to be in production within a year [2]. This migration toward data mining implies that maintaining an adequate warehouse may soon become simply the price of entry, rather than a competitive weapon. Data warehouses can increase revenues (by effectively targeting customers) or decrease costs by screening or discouraging customers who are most likely to churn or otherwise be unprofitable. A data mining project at US West detected that customers want nonprice features such as technical support or free service for new features; this discovery enabled the company to preserve revenues. Besides US West, Bell Atlantic and Alltel claim that data mining is producing results. Indeed, telecommunications service providers enjoying large benefits from their data warehouses would be shrewd not to publicize their successes.

10.3 Organizational structure

Telecommunications service providers will need to build organizational structures that support, rather than obstruct, their business objectives. Incumbents and entrepreneurs both will need to transform their organizations. Incumbents and the largest providers will need to reduce their organizational costs and direct their organizational priorities toward the customer. New entrants will need to grow in ways that increase their organizational costs at a slower rate than their organizational size.

The traditional monopoly telecommunications provider was usually completely vertically integrated. The monopolist built and operated the network, conducted all the marketing and back office operations, and in many cases owned the manufacturing facilities. At AT&T's divestiture of its operating companies, the network was divided, as were the supporting R&D outfits (in Bellcore and Western Electric), but not the business functions or the markets served. AT&T and the operating companies decisively reduced the number of management levels in each organization. Most of the operating companies reorganized their operations into customer segments and product lines. Today's telecommunications organization produces more operating capacity with decidedly fewer organizational resources. Ameritech started to reorganize its business along business unit lines in the early 1990s and simultaneously focused its resources regionally to become more customer-focused.

New entrants have an opposite danger; growth in competitive markets is so high that new service providers will constantly be building on the organizational infrastructure. The danger is that companies expecting significant growth can overextend their management ranks, either in preparation for the addition of staff, or simply to reward an employee for a job well done. One mid-sized telecommunications provider in a rapid growth phase presented its strategic planning committee with an organization chart that featured 10 levels of management, the same number of levels AT&T maintained prior to divestiture.

Merging companies compound quandaries of organizational structure. The separate companies often do not start out as efficient as they can be; combining them duplicates efforts, especially in the management ranks. Furthermore, many companies justify the high acquisition premiums they pay by asserting that organizational efficiencies will lower costs. At best, the cost reductions associated with redesigned business processes and the integration of information systems will take longer than management desires. At worst, a third layer of process overlays and reconciles conflicts between the existing two, adding to overall cost.

Telecommunications service providers acknowledge that the changes of the past decade will not satisfy the requirements of the competitive environment. An Andersen Consulting survey of senior executives at telecommunications providers worldwide determined that 80% of executives anticipate altering their organizational structures in the next 10 years [3]. The respondents predicted that telecommunications service providers would spin off divisions to create a more entrepreneurial and competitive climate. The same study predicted three emerging organizational models: dominating one area of telecommunications; leading the market in all areas of telecommunications; or acting as a solutions provider, bundling services from a variety of providers to meet the individual customer's need.

The remaining organizational incongruity among incumbent service providers is their continued determination to operate in both the wholesale and retail markets. Regulators, attempting to create an interim source of competition until facilities-based carriers were widespread, imposed wholesale roles on RBOCs. The practice of providing wholesale and retail services concurrently is not limited to RBOCs and is indeed common among service providers as a matter of strategic choice. There are many reasons that a wholesale and retail model will be an unsuccessful strategy for incumbent providers in the long term. IXCs have been successful at performing wholesale functions only because it represents a tiny part of their business. Incumbent local providers have already learned that their wholesale customers find it difficult to assume that their incumbent suppliers want them to be successful. Employees of incumbent providers are understandably threatened by the competition. When opportunities to serve one's own employer conflict with serving a competitor, errors of omission or commission are possible. Furthermore, a telecommunications service provider simply cannot be a superior performer in disparate markets with disparate needs. Several telecommunications providers attempted to separate their wholesale and retail businesses with structural separation. One was Frontier (formerly local provider Rochester Tel) and the other was SNET. Neither of these providers was under any regulatory requirements to separate the two businesses. In fact, both were unaffected by the consent decree that prohibited Bell companies from offering local and long-distance service, so they were each integrated providers when they made the structural separation. Ironically, each has been acquired, SNET by Southwestern Bell and Frontier by Global Crossing, which was until the acquisition a wholesale-only provider. While other incumbent LECs in the United States have created wholesale divisions, none has established a separate financial operation. One incumbent local service provider devised transfer costs and procedures for the divisions to work together but would not take the bold step of allowing the retail operation to purchase services from any other wholesaler. Later, British Telecom announced plans to divide its wholesale and retail operations but stopped short of considering a total divestiture for either of the new entities.

Competitors have petitioned to create structural separation for carriers struggling with channel conflict. One IXC petitioned the FCC to require that local service providers create distinct wholesale and retail entities. While the FCC has been willing to impose conditions on lifting the restriction against long distance, the structural separation never became a serious consideration. One residual problem is that the placement of assets and costs affect the wholesale prices, which remain under regulation. Still, the experiences of Frontier and SNET offer some hope that integrated telecommunications service providers in the United States will see it in their own business interests to divide their operations into wholesale and retail entities.

10.4 Reengineering

Business process reengineering involves dismantling complex business functions and recreating them in slimmer, more elegant ways. Reengineering projects promised to reduce workload dramatically and utilize information technology as an enabler of change and process improvement. The seminal reengineering works were first a *Harvard Business Review* article and later a book by Michael Hammer and James Champy [4]. The first five years of reengineering generated a significant amount of publicity and highly publicized failures. Notwithstanding many excessive predictions about its benefits and the number of successful projects, the business community has positioned reengineering into a more practical application after a few years of practice. Information systems such as ERP software have enabled the newest reengineering efforts, and most recent process redesign efforts are more limited and more controlled than the enterprise-wide reengineering mania of its first few years.

Reengineering captured the attention of many telecommunications service providers in the last decade. Many service providers downsized their work forces in anticipation of competition. The cost savings took years to materialize, but most telecommunications service providers are leaner and more prepared than ever to compete with the efficient new entrants to the industry.

Telecommunications service providers suffered setbacks during their reengineering efforts. First, the traditional human resources policies were quite paternalistic and at odds with work force reduction, including those that enticed employees to leave with financial and other incentives. Moreover, reengineering efforts that succeed generally do so because they reduce the work load at the same level as the reduction in work force. Many telecommunications service providers admitted publicly that their reengineering targets were too aggressive and that service quality suffered. US West underwent a reengineering effort to eliminate layers of management in the customer service area, reducing its work force by almost 17% [5]. The company's failure to anticipate a surge in demand for new lines caused delays in installations and answering customer service and repair calls. MCI acknowledged that three of its own reengineering efforts ended in failure. AT&T underwent a reduction in its work force by nearly 20,000 employees through attrition and generous buyout offers. Indeed, the incentives (along with a growing industry and a booming stock market) were so popular that AT&T met its downsizing objective about a year early. While buyouts reduce the amount of criticism a service provider will receive for the work force reduction, voluntary programs can risk losing some of the best employees, because the most desirable employees can find employment elsewhere. When WorldCom acquired MCI, it needed to increase its offer to counter a competing bid and justified the increase by reporting anticipated additional savings by $20 billion. These savings would be achievable only if the merged company were as successful at integrating its systems as WorldCom had been in the past through its other acquisition experiences. Because MCI's systems and processes were undoubtedly more complex than those of earlier acquisition candidates, the newest integration was undoubtedly the most challenging to date. WorldCom's proposed merger with Sprint would have complicated the integration with MCI, as the MCI integration would probably not have been completed by the time the newest integration with Sprint was necessary.

Intermedia Communications, a facilities-based local service provider targeting the business community, underwent a successful reengineering effort to meet its stringent standards for system availability [6]. The high standard the service provider set for itself, coupled with its rapid growth, fueled the company's concern that it needed to streamline its business processes and especially to reduce its mean time to repair (MTTR). Within three months, the first metrics showed improvement, and eventually the service provider exceeded its reengineering objective.

10.5 Skills and training

Telecommunications service providers, recognizing the contributions of a skilled work force, have increased their expenditures for training and skill development in response to the competitive market. According to the American Society for Training and Development, service providers spent nearly $900 per employee in training in 1997, and Bell Atlantic reported a 10% increase in 1998 over its already considerable training budget of the previous year [7]. Training budgets for telecommunications service providers exceed those for other industries by nearly 30%.

One reason that training costs are rising is that telecommunications service providers train employees in areas that are rising in importance in the competitive market. While technical training was always required, a wider audience for technical training includes the direct sales force (as services become more complicated than dial tone). The decentralization of decision-making and the necessity for employees to work in outwardly focused teams has created a need for diversity training and for management training. Telecommunications service providers that aim to differentiate based on customer service have created programs to improve service representative performance, with sales training included to increase the potential for cross-sales.

Service providers realize that a trained work force can help to retain customers and that a commitment to recurring training can retain valued employees. The rapidly changing technologies of both telecommunications and information systems creates a shortage of skilled workers and demands that existing staff skills be upgraded more frequently than in other industries. Training telecommunications staff has become a valuable market for support organizations in the industry.

Companies that base their differentiation upon specific core competencies should invest disproportionately in the training that makes the competence possible. For example, a company that differentiates on its provisioning excellence needs to ensure that all participants in the provisioning process, from service representatives to technicians to billing personnel, understand the process and are skilled in its execution. Customer-facing staff also requires sales training and training in customer service.

Another way to upgrade the quality of the technical and professional staff is to invest in recruiting as well as training. Some telecommunications service providers such as AT&T and GTE have joined a consortium of 14 companies called the Talent Alliance. One of its objectives is to codevelop and share training programs. Another benefit of the alliance is that members can use each other's key workers for projects, reducing the recruiting and training needs of members.

All of the RBOCs recruit workers both internally and externally, and growth in the industry, coupled with a tight labor market, has created recruiting challenges. Demand is highest for engineering, information technology, and customer service positions, reflecting the areas in which telecommunications service providers will position themselves uniquely. Recruiting strategies include starting bonuses, retention bonuses, Internet-based recruiting, employee referrals, community relationships, contract recruiters, advertising, and sponsorships of job fairs. This represents a large cultural change for these service providers, which historically promoted from within and hired generally only at the lowest rungs of the corporate ladder.

Bell Atlantic's Next Step program uses classroom-based training using instructors from the company and from vendors [8]. Partnerships with local community colleges will enable the service provider to expand training options, and employees will be able to obtain two-year degrees for their efforts. Ameritech works with colleges and offers direction for the program's curriculum.

10.6 Supporting tools

Many of the organizational advancements of the last decade would not have been possible without the commensurate advancements in information technologies. Communications technologies eliminated geographical borders and organizational boundaries. Information technologies have made it possible to decentralize decision-making and analysis to virtually any level of the organization.

Groupware

Groupware is collaborative software that enables workers to communicate freely from multiple locations or even multiple desks within a single office. The first groupware software was Lotus Notes, later supported by its Domino intranet server. Microsoft's Exchange offers groupware capabilities. US West has conducted a joint marketing effort with Lotus, now a division of IBM. MCI, BellSouth (in partnership with EDS), GTE, and AT&T are among the service providers that have offered groupware solutions to their customers.

The sophistication and affordability of computing power for groupware have facilitated the emergence of knowledge management. A corporate intranet is vital for the successful implementation of knowledge management. MCI was among the first telecommunications service providers to establish a library on its internal corporate network. Its library hosts tens of thousands of visitor sessions every day and provides knowledge from corporate communications, competitor activities, news, market research, and selected Internet sites. The intranet is directed at all MCI employees, not just the technical staff, and has increased in sophistication, including its search capabilities.

AT&T supports an intranet that offers knowledge to its employees and contains a database of contacts for further inquiries. Because intranets are so dependent on communications solutions, telecommunications service providers that stay ahead of the technology are able to create customer solutions from their proprietary systems. AT&T sells its knowledge management solution and identifies eight process steps to assist its customers in implementation and use: analysis, assess, design, build, pilot, deploy, operate, and maintain.

Service providers use groupware extensively for communication, calendaring, and other functions most common to groupware applications. Future requirements will increase the inevitability of groupware. Telecommunications service providers use groupware internally to meet their growing collaboration and project management requirements in an increasingly global corporate footprint. The tool can enable service providers to offer distance learning to their employees and potentially to suppliers and customers.

Self-directed teams

In the last decade, organizations have restructured to meet rapidly changing marketplace requirements. Among the changes are flatter, horizontal structures replacing the traditional vertical hierarchy. This reduces costs and facilitates productivity. Management has empowered teams, organized around business processes, to make day-to-day decisions. The resulting teams are cross-trained and highly skilled; high performance produces rewards for team members. While this approach is a departure from tradition in many companies, it is especially new for the oldest telecommunications providers. In the interest of competitive success, management has embraced the new organizations.

According to the ERIC Clearinghouse, a national education information network, AT&T reorganized its Network Services Division around processes and awards bonuses based on customer evaluations. GTE redesigned its customer contact process to offer a single contact linked through one telephone number, including sales, billing, and repair. Bell Atlantic can fill within hours some service orders that once took 15–25 days.

GTE Directories has used a variety of self-directed teams for process management, quality improvement, and cross-functional leadership. The company's team management efforts contributed to its receipt of a Malcolm Baldrige Quality Award.

AT&T met with its largest union to define the workplace of the future. The corroboration between the service provider and union in the planning process is by itself a departure from traditional corporate-union relations. The resulting organizational philosophies reflect the need for management and labor to work together to meet customer needs in an intensely competitive environment.

10.7 Role of information technology

Information technology is the enabler of organizational effectiveness initiatives. It is not a coincidence that the emphasis on new organizational structures and processes happened when personal computing and the Internet became ubiquitous. Knowledge management arose when information and communications technologies could finally support the large amount of data transfer required for its success. Without intranets, knowledge management is too weak to justify its investment. Horizontal organizational structures are based on defined business processes. To eliminate the requirement for vertical (supervisory) intervention in business processes, information technology support must be able to provide the breadth of information formerly contributed by middle management.

Even the move away from vertical integration relies on technology to succeed. When telecommunications service providers eliminate their traditional vertical integration, they need to sustain communications with their suppliers at the same quality as their former internal paths. Extranets and associated security measures (including emerging business-to-business e-commerce) depend on reliable and affordable technology.

All reengineering initiatives have information technology at their core. Distance learning places communications and information technology at the center of the training function. Use of information technology in business process redesign can eliminate the need for training in the first place. Data CLEC Covad automated its supply chain by creating a flow-through provisioning system. Among its many advantages is that the reengineered and fully automated process eliminates the need for training service personnel to make decisions or rekey information delivered by the software. Technology can eliminate errors and obviate training in the service provider call center environment as well, by using scripts that execute based on customer responses. Eventually, these scripts will lead to total self-provisioning for customers that choose to use Web-based customer service.

References

[1] Bane, P. W., et al., "International Companies: Benchmarking Performance," *Telecommunications*, Vol. 29, No. 2, p. 21.

[2] Handen, L., and P. Boyle, "Data Warehousing: Managing Knowledge," *Telecommunications, International Edition*, Vol. 32, No. 1.

[3] "Change—or Die," *America's Network Telecom Investor Supplement*, Vol. 103, No. 15, pp. 6–7.

[4] Hammer, M., and J. Champy, *Reengineering the Corporation: A Manifesto for Business Revolution*, New York, NY: Harper Business, 1994.

[5] Lawyer, G., "Leaner, Meaner—and Busier," *tele.com*, Vol. 3, No. 4.

[6] Levine, S., "Keeping the Bar High," *America's Network*, Vol. 103, No. 7, p. 56.

[7] Turner, T., "TeleCom 101," *tele.com*, Vol. 4, No. 6, p. 51.

[8] Schmelling, S., "Reel 'Em In," *Telephony*, Vol. 236, No. 3, p. 22.

11
Competitive Parity

We can hope for competition all we want, but that doesn't mean it's going to happen.

—Former FCC chairman Reed Hundt

In competitive markets, the definition and positioning of any telecommunications service provider will develop not only through one's own characteristics and actions but also through those of its competitors. The way that service providers learn to anticipate competitor activities, lead, follow, or selectively ignore, and outmaneuver competitor actions in the marketplace can affect the service provider's success as much as most internally driven strategic decisions and actions.

11.1 The nature of competitive parity

Competitive parity can be defined as the strategies undertaken by a telecommunications service provider to maintain its relative stature in its brand, pricing, market scale, or other defining characteristics in comparison to its competitive surroundings. Competition, by definition, is the rivalry among industry participants to attain a limited reward, whether it is customers, revenues, licenses, or any other prize. Effective competition exists when competitors are sufficiently matched that they can apply significant mutual pressure.

Parity issues arise primarily between competitors, but the same principles apply to other external relationships maintained by telecommunications service providers. Virtually every relationship requires negotiation, and each successful negotiation concludes with an allocation of value between the two negotiating entities. Negotiations between telecommunications service providers and suppliers, or customers, or regulatory bodies all require constant repositioning based on the completed or anticipated actions of the other.

Recognizing the role of competitive parity in the telecommunications industry, at least one leading service provider has maintained a senior management position that includes the concept of "competitive response" in its title. Generally, though, service providers act with recognition of the role of competitive parity, but they do not isolate its activities as a separate corporate function. Furthermore, incumbent service providers tend to be reactive rather than proactive during the transformation to a competitive mind-set. Market researcher Frost and Sullivan noted that direct-to-home (DTH) satellite service in the United States boosted its market share from 5% to 12% between its launch in 1995 and 1998. Still, cable companies did not upgrade their systems very aggressively and limited their trials to areas that sustained a high subscriber loss to DTH.

Many incumbent local telecommunications providers in the United States accelerated their DSL service launches when cable modems began to get a foothold in the consumer market. Indeed, telecommunications service providers frequently sustain criticism for delaying product launches until competitive entry is at hand. Only the service providers involved know whether the timing of their DSL launches was strategic or simply a market reaction. Whether an earlier launch would have offered competitive advantage over cable modems can only be hypothesized. DSL, because of its integration with installed facilities, is likely to be more forgiving of a late entry than future technologies. Nonetheless, cable modems continue to lead DSL in consumer penetration. Insight Research estimated that by the end of 2000, cable modems would exceed DSL access by 2 million cable subscribers versus 1.1 million subscribers to DSL service.

Businesses in all industries have borrowed liberally from military terminology and strategies to succeed in intensely competitive markets. MCI's Friends and Family program was a frontal attack on AT&T's strength in market share. Sprint's "pin-drop" campaign is another attack on AT&T's strength; this time the target is its high-quality network. AT&T's competitive response, its "True Voice" brand, could not surpass the strength of Sprint's focus. Attacks on weaknesses are often successful as well.

Many new entrants have profitably created markets in areas that simply could not command sufficient attention from incumbent management. The public telecommunications segment (pay phones) was highly regulated for incumbents, even while new entrants could select the best markets to serve. While new entrants had their own challenges because their suppliers maintained channel conflict, the incumbents' portfolio of telephone locations included, by regulatory requirement, many very unprofitable sites. The line of business simply could not warrant sufficient management attention through its size or its profitability compared to other divisions. To some extent, even the highly profitable directory segment is similarly challenged because of its size.

One military strategy, launching offensives on many fronts, is practiced primarily among the giants. Its disadvantage in the telecommunications market is that few of the industry leaders have so many resources and such targeted options as to be able to overpower their competitors by overwhelming them with investment. Wisely, the largest service providers are investing in the growth of the industry rather than capturing additional market share. If they chose to do so, industry leaders could overpower smaller entrants with offensives on several fronts, causing them to splinter their attention and resources. Fears of legal reprisals and the temptation simply to merge with attractive competitors will probably prevent the giants from utilizing this strategy. GTE conducted an end-run offensive, maneuvering around competitor markets, when it purchased BBN Planet, an ISP and backbone data network, and entered a strategically important market well before its larger competitors. Another strategy borrowed from warfare is the guerrilla offense, targeting a small market niche in which conditions are most favorable to the service provider. This is an especially useful strategy for small providers, because it creates an opportunity to claim ownership of a market segment and maintain competitive advantage. Cellular providers VoiceStream and Omnipoint chose to use GSM technology as a differentiator in the U.S. mobile market. For the segment of customers who believe the GSM protocol to be superior and those who need to roam in Europe, where GSM is the wireless standard, this offers competitive advantage. The merger of the two providers will enhance the market size of the provider and its wireless footprint in the United States. In its first years, MCI used the guerrilla strategy in its efforts to gain competitive advantage over AT&T. The upstart provider's constant legal actions against the IXC, discounted prices, and intermittent promotional activities were classic guerrilla tactics. Another military strategy involves making preemptive strikes to secure an advantageous position that competitors cannot match. While the AOL merger with Time Warner has many strategic advantages, the merged company's preemptive benefit is that each is the uncontested leader in its own market. It would be very difficult for other providers to replicate the merged entity through alliances or growth.

While new entrants will undoubtedly embrace these offensive strategies to enter the telecommunications markets, the incumbent carriers primarily employ defensive strategies to protect the markets they serve. Defensive strategies comprise activities that reduce the likelihood of attack from others, minimize the impact when attacks do occur, redirect challengers to other rivals, and strengthen the present competitive position. The most obvious example of incumbent providers trying to reduce competitive inroads is their constant journeys into the courtroom. RBOCs have balked at most efforts to open their networks to competitors, and AT&T operated exactly the same way when its own cable distribution was threatened with "open access" to competitors. Incumbent providers generally act defensively in their introduction of new technologies, probably less a competitive strategy and due more to tradition and inertia. As markets become more competitive, telecommunications service providers quickly adapt to the requirements of the marketplace. SBC launched Project Pronto as an effort to protect its competitive position with its customer base, vulnerable to other broadband access providers. The project, estimated at $6 billion, improves high-speed access through DSL and other broadband technologies. Generally, though, RBOCs have demonstrated procrastination in deploying DSL. Their typical deployment time frames imply that their strategy is driven defensively or that the delays are due to valid fears that affordable DSL will cannibalize other, more profitable service lines such as ISDN in consumer and small business markets. Another defensive strategy is to broaden product lines to cover gaps that competitors could exploit. The extensive but understated trials of IP telephony services among facilities-based providers are an example of this defensive approach.

11.2 First-mover advantage

The timing of a strategic move is often as important as the nature of the move. First movers into a technology market can gain advantage by deciding which technology to use, when to introduce services, and initial price levels [1]. In high-speed Internet access, the first service provider in a market, whether it uses cable, DSL, or other technology, sets the price and package competitors have to surpass. Early adopters will subscribe to the first offer; the second tier of customers will be harder to convince. McKinsey's research into early broadband pricing suggested that providers were charging less than the break-even price for access. McKinsey concluded that the service providers' intention was to capture the first mover position in their markets, planning to recover profits by retaining customers at the same levels as technology prices dropped. Additionally, first movers derive cost benefits when economies of scale are present. Being the first to market also enhances the image of the provider, a critical branding strategy for service providers desiring to differentiate their services based on superior technology. Sprint's reputation for excellent network quality is partly based on its buildout of the nation's first fiber network and partly because of exceptional image advertising and branding.

Providers of conventional cellular services were able to achieve higher profits on their services than the present set of competitors. Though the service was much more expensive to provide, the first-mover advantage (and the duopoly nature of the market) enabled the rivals to build more profit and fewer features into their offerings and capture the backlog of customer demand.

First movers can take advantage of favorable relationships with suppliers when the market potential is unknown and they have their choice of distribution channels, especially when there is pent-up demand for the new service. Being the first mover can hold several disadvantages as well: The costs of being first are high in any industry and are especially high in technology markets. Rapid technological change can enable followers to capture mass-market share instead of the small early adopters, and later customers are available when the technology is less expensive. Followers can inexpensively copy many skills and business processes honed through the trial and error of first movers. Those benefiting from first-mover advantage need to be vigilant about competitors closing the gap. Active competitors attack their own competitive advantages as decisively as they would any competitor's.

In technology markets, a second-mover advantage can reduce or eliminate first-mover advantage. The first mover incurs the cost of educating the market about the benefits of the new service, driving the risks down for later entrants. The first mover creates the critical mass needed to create market activity, which benefits the second mover without cost. In technology markets, first movers often need to invest in expensive and first-run technology that can become obsolete before the second entrant enters the market.

New fiber optic networks constitute a significant competitive threat to the incumbent carriers in the United States. Qwest, Level 3, Williams, and Broadwing (formerly IXC) have added considerable backbone capacity in the interexchange market. According to the Yankee Group, the Qwest network in 2000 contained more capacity than the networks of AT&T, MCI WorldCom, and Sprint combined. Level 3's planned network is expected to eclipse the capacity of Qwest and the leading IXCs. This is a significant strategic threat to the incumbent IXCs, which are responding by upgrading their own networks. The long-distance entry of the RBOCs will encourage the service providers with nationwide or near-nationwide coverage (most likely SBC and Bell Atlantic) to build their own backbone networks as soon as restrictions are lifted sufficiently to make them profitable.

Whether RBOCs or wireless carriers with nationwide footprints should build backbone networks is a difficult strategic decision. Parity pressures will encourage them to do so to maintain their standing against their competitors. On the other hand, the backbone networks are already at overcapacity, and they are likely to remain so for quite some time. Vertical integration is not always the proper action, even when competitive parity is tempting. Backbone facilities are the likeliest of all network components to remain an undifferentiated commodity in a competitive market. Opportunities for differentiation, such as superior network management or exceptional transmission quality, would work in a wholesale business model, but are not as viable for a retail service provider. Unless there is a significant cost advantage to operating these networks in-house, service providers can direct their investment more successfully to differentiating services in some other domain.

11.3 Game theory

Game theory is the mathematical theory of bargaining. It was developed in the mid-twentieth century by mathematicians John Von Newmann and

Oskar Morgenstern and updated in the work of John Nash. Many other disciplines, including the business community, embraced game theory as a way to make optimal decisions when they involve several active participants and inadequate information. The newest models view competition as a process of strategic decision-making that is performed under uncertainty. It is relevant to competitive parity because it emphasizes the actions and reactions of competitors when a service provider makes a strategic decision or action. The most famous example of game theory is the prisoner's dilemma, in which police separate two criminal suspects and provide each with a matrix of choices for confessing (defection against the partner) or not confessing (cooperation with the partner) to the crime. Suspects are not aware of their partner's decision whether or not to confess before they make their own choices. Each suspect receives the minimum sentence when he or she confesses and the other does not. Each suspect receives the maximum sentence when he or she does not confess but the other does. If both confess, they both receive a very high sentence; if neither confesses, they both get a moderate sentence. The extreme sentences occur when partners do not agree on whether to confess, but a partner who maintains innocence if the other confesses gets the worst outcome. The reason that the prisoner's dilemma has captured the interest of economists and the business community is that it covers general strategic principles. It encompasses tradeoffs between selfish goals and the common good, demonstrates the ability to cheat in a cartel, and provides insight if a decision- maker does not know how one's competitors will act or respond to one's own actions. Another feature of the prisoner's dilemma is that the game changes significantly if the players repeat the game. As service providers can be relatively certain that they will continue to interact with their competitors in the future, the game becomes more relevant and more useful.

In a duopoly or oligopoly, virtually every strategic decision is a candidate for assessment through the viewpoint of game theory. Game theory applies to price changes, because one service provider's decision to raise or lower rates will affect customer demand and competitor actions. It can also apply to new buildouts, acquisition initiatives, or spin-offs of underperforming businesses. Virtually every significant strategic decision has competitive repercussions and potential competitor retaliation. The Nash equilibrium occurs when a participant's strategy is optimal in light of the potential strategies of others. In the prisoner's dilemma, confessing is the best strategy. This applies in competitive situations as well. The Nash equilibrium occurs in the highly elastic pricing of telecommunications services. Figure 11.1 demonstrates the decision process for a telecommunications provider deciding to lower prices or to respond to a competitive rate reduction. Absent other considerations (and there are always other considerations), the optimal strategy is to match competitor price decreases.

Game theory also applies to advertising, in that telecommunications service providers need to advertise heavily when their competitors do so. If no telecommunications service provider were to advertise at all, each competitor would benefit. Nonetheless, agreeing not to advertise is similar to participating in a cartel. Any competitor that breaks the agreement and advertises will benefit greatly. In fact, game theory is widely applicable to all business decisions concerning expenditures of discretionary size, such as community support and campaign contributions.

Not every situation includes a dominant strategy that makes the decision apparent without knowing the competitor's behavior. In the pricing example shown in Figure 11.1, My Telco's strategy should always be to lower prices to the minimum, whether or not the competitor has acted. In other situations, the optimal outcome depends on the competitor's

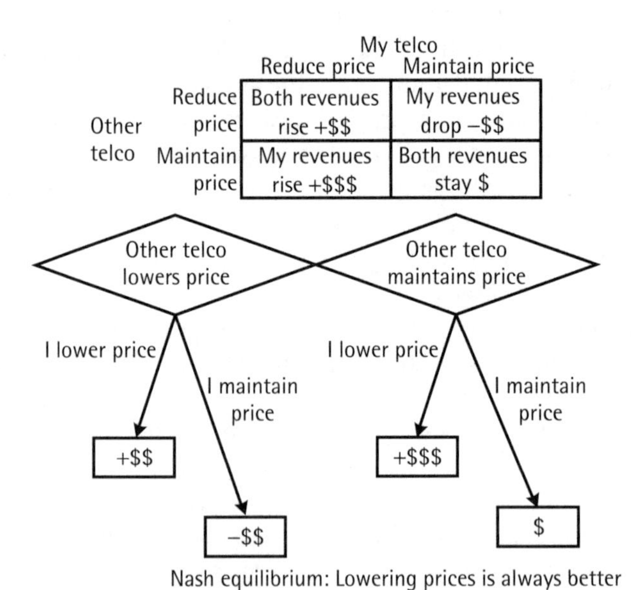

Figure 11.1 Prisoner's dilemma and pricing strategies.

ultimate action. In addition, the outcomes for cooperation and defection change if the participants play the game more than once.

The most familiar example of the use of game theory in the U.S. telecommunications industry is in the bidding for wireless PCS licenses from the FCC. The FCC used game theory experts to construct the auction to maximize revenues quickly. Auction participants engaged their own game theory specialists to assist them in bidding. As the structure was a multiround auction, participants were able to learn from earlier bids, in a way that would help them make later choices. Companies used other techniques to disguise their intentions, such as bidding beyond their eligibility and unexpected exploratory bids in markets that were not apparently of interest to them. The auction process was of value to the FCC and other government bodies. The knowledge gained from the license auction will undoubtedly apply to future auctions of spectrum and other government resources such as rights to natural resources, Treasury bills, and permits.

In the United Kingdom, a spectrum auction round for universal mobile telecommunications system (UMTS) licenses used game theory on both sides of the auction. This technology, the foundation for 3G wireless services, is widely anticipated but highly uncertain. Brazil, India, and Japan are among the countries where government auctions returned higher-than-forecast or other surprising results.

11.4 Choosing parity battles

The concept of value innovation, described in Chapter 8, identifies competitive strategies that depend more on internal capabilities than outside influences. Focusing on competitors excessively can result in reactive, imitative, or incremental actions. The highest performing companies focused instead on expanding existing markets or creating new ones.

In rapidly growing industries such as telecommunications, service providers will have many opportunities to expand their markets to meet their own core competencies instead of zero-sum battling for the same customer dollar. Value innovation is not the same as technology innovation. Lotus Development (now part of IBM) created a significant technology innovation when it introduced Lotus Notes, the first significant groupware application. This application threatened Microsoft's operating system market and in particular its emerging Exchange Server product. In response, Microsoft continued its development of the competing product, but it also Web-enabled its office productivity software, unleashing the Internet market. The variety of opportunities that the Internet has added to Microsoft's portfolio certainly eclipsed any potential market loss that it sustained to Lotus.

Telecommunications service providers need to recognize whether their innovations and differentiators constitute rival goods or nonrival goods. When one service provider uses a rival good, such as a proprietary technology or a superior-performing salesperson, that resource is not available to its competitors. DirecTV, the satellite television provider, offers a package of its television service and broadband Internet access, which it calls DirecPC. This represents a rival good and a competitive differentiator, because competitors could not copy the service with their own technology or replicate the bundled price. On the other hand, when one service provider offers frequent-flier airline miles to its customers as an incentive for using its network, that is a nonrival good. One service provider's offer does not preclude other service providers from following suit. Competitive imitation of this kind is flattering, but it eliminates the competitive advantage of the service provider that offered it first. For example, all of the major long-distance carriers in the United States offer airline miles as an incentive to presubscribe to their service. Offering miles has become a cost of customer acquisition.

Moreover, the first service provider might have taken a financial risk and certainly took a marketing risk by initiating the nonrival program. If it is successful, others can copy it at almost no cost and almost no market risk. In service industries, this challenge is especially difficult. A service provider can patent its technology, if it can find one that truly differentiates its service. Patents do not protect many of the other service differentiators. The company with on-line billing maintains an advantage only until a competitor can develop a similar infrastructure. In fact, the competitor has the opportunity to review the on-line billing operation and improve it for its own use. What the first competitor does have is first-mover advantage and time to develop the next innovation to retain its customers while its competitors are catching up with their first one.

11.5 Retaining the customer

Switching costs are the economic cost of changing service providers, some of which are borne by the customer and some by the provider. When interexchange presubscription was introduced in the United States, the local service provider added a small administrative charge to the customer's bill for changing providers. The charge might have constituted a switching cost if long-distance providers did not routinely cover the charge as a cost of customer acquisition. For the largest customers and some wireless customers, there are penalties for terminating contracts before their term is completed. Another switching cost for wireless customers is the lack of ability to apply the investment in the telephone itself to a new service provider, such as when moving to a new wireless provider with a different technology. When hardware is involved, switching costs are often quite large. Ironically, hardware standards, which often help greatly to create viable markets, tend to eliminate switching costs for those holding proprietary technology. Still, service providers benefit as much as they suffer when switching costs are eliminated, other than incumbent providers.

For consumers, promotions often serve as a negative switching cost. If an ISP offers a free month of service or an interexchange provider offers free minutes or, in some cases, an outright payment for switching, the customer benefits economically from switching. Nonetheless, not every switching cost is purely financial. While it is very easy to switch from one ISP to another, and churn in that market is quite high, one significant switching cost has probably kept churn from being even higher. The customer's e-mail address contains the domain name of the ISP, and the customer's access to that address will expire if the customer leaves. ISPs will not forward mail on behalf of former subscribers. A niche market has emerged to give lifetime addresses to customers; many portal sites are happy to give customers an address and a reason to visit the site on a daily basis. Microsoft's Hotmail service chose Web-based e-mail as its primary business model and is now among the most popular sites on the Web. While the mail service is free, banners paid for by advertisers surround the message retrieval and management screens. At the end of the Hotmail session, Hotmail sends the user to the Microsoft Network portal with a new set of banner advertising and shopping opportunities.

AOL has included many other switching costs into its infrastructure. One feature of the service is the subscriber's ability to manage one's calendar on-line. Friends can review the calendar and set dates against it. The subscriber contemplating a switch will need to recreate the entire calendar and notify all contacts when embarking on a new ISP. Again, portals have created their own calendar features to capture customers who have not committed to AOL. Other switching costs accrue from AOL's large market share. Buddy lists and instant messaging enable users to detect when specified other subscribers are on-line; most other services do not offer this feature. AOL has also developed software that enables nonsubscribers to appear on the buddy lists of its members.

Commodity markets create intense competition and narrow margins. Competitive telecommunications service providers that find ways to lock in customers will command higher prices and stronger customer retention. One of the largest competitive challenges in the consumer telecommunications market is the freedom that customers have to switch providers. In an effort to ensure a level playing field, regulators are doing whatever they can to create as few switching costs for customers as possible. Carrier preselection enables customers to choose their long-distance provider, and competitors ensure that changing providers is quite easy. Local number portability, which represents the telephone equivalent of removing the e-mail address switching cost, allows customers to keep their telephone numbers when they switch local service providers. Number portability is a regulatory requirement in most competitive markets.

Switching costs can affect the customer and the service provider differently. If a change in long-distance carriers incurs an administrative charge to the local exchange company of $5, the charge is either $5 to the customer or $5 to the new long-distance provider. If the service provider offers a $100 check in exchange for switching, the value of the benefit to the customer or the cost to the carrier is equal. If the service provider instead offers 2,000 free minutes over the next two years, the cost to the provider is decidedly lower than the value to the customer. The customer benefit is at retail, and the carrier's outlay is at cost. Furthermore, the amortization of the minutes over time takes advantage of the time value of money and the likelihood that per-minute prices will drop further, and any potential customer churn before the minutes have been expended, while not desirable, at least consumes only part of the acquisition cost.

Eliminating switching costs is simply the price of acquiring customers, and the amount is generally straightforward to calculate. The more challenging strategic effort is to construct a customer acquisition program in which the switching costs borne by the customer are simply unassailable by competitors. This is the link to competitive parity. The example described earlier, MCI's Friends and Family program, created significant switching costs for customers. In the case of Friends and Family, the terminating customer loses a discount and eliminates the discounts enjoyed by others.

Furthermore, the switching cost could not be reversed by any competitor, and especially not by AT&T. The same holds for AT&T's Digital One Rate program. When it began, customers requiring affordable nationwide access could not be tempted by competitive offerings.

Loyalty programs represent a switching cost created by the service provider to compensate for the lack of inherent differences or switching costs in the service itself. While several telecommunications service providers have offered loyalty programs that increase the rate of awards with the level of customer expenditure, few have attempted to retain customers by increasing the rate of awards through the customer's longevity. An example of the former is to provide one airline mile for each dollar of usage. An example of the latter is to offer one airline mile per dollar the first year and two miles per dollar the second year. The advantage of increasing rewards for longevity is that it matches the service provider's outlay to the benefit. In addition, once the customer has achieved a certain time with the provider, competitive retaliation is impossible.

Another opportunity for telecommunications service providers to retain customers through lock-in is to communicate frequently to them about plans for service upgrades, promotions, or price reductions. The opportunities to capture them for a new term range from preconstruction sign-up for emerging services such as DSL before they become available or signing a new contract prior to the end of the contract term. Some wireless carriers have been successful at reducing churn by identifying customers, using a data warehouse, in the last month of a wireless contract. By offering an attractive rate for a second year of service, wireless service providers have been able to eliminate the customer's search for an alternative provider.

One caution for telecommunications service providers is that locking in customers can draw regulatory scrutiny. One RBOC discontinued its escalating awards program after competitors complained about a potential abuse of monopoly power. The line between intelligent marketing and monopolistic practices was at the center of the Microsoft antitrust case.

11.6 Raising the bar

Competitive battles tend to escalate over time. Service providers jockeying to gain market advantage are in a perpetual struggle to maintain their lead or retake the lead when someone else jumps ahead. The result is that the benchmark of necessary performance tends to move upward over time. Competitive escalation is most evident in pricing, in which one company takes a lead and others choose to follow or modify the new pricing package or ignore the competitive action. Whether industry leaders choose to react to price changes often depends on the source of the new pricing. Internet-based long-distance providers often advertise prices well below those of established carriers. Furthermore, IP-based providers have pioneered monthly flat-rate long-distance service, which service provider marketers cannot ignore. The largest providers generally ignore these offers in the short term and fold them into longer-term strategies. Among the larger service providers, Sprint has consistently acted as the pricing leader: Sprint introduced flat, per-minute pricing in the mid-1990s. Sprint is often the first to drop per-minute prices, was the first to offer a flat-rate bundle of long-distance minutes on weekends for an inclusive price, and has led the leading service providers in promotional pricing. In its third-place provider position, it is expected that Sprint will take some initiative, but the company exceeded expectations and influenced the industry in excess of its small market share. AT&T's pricing strategy is to take its cues from competitor price reductions but utilize its strong brand and cater to its customer base by simplifying the price structure instead of the two-tiered packages generally offered by Sprint and MCI WorldCom. In this manner, AT&T can command a slightly higher per-minute rate.

Pricing is not the only area of importance in competitive parity. The scope and scale of the business are acceptable only in comparison to the attributes of competitors. Companies that do not act aggressively fall behind those that do. In delaying its nationwide wireless strategy, MCI WorldCom eventually forced itself into a near-desperate position to buy a wireless provider. By failing to grow significantly through assertive acquisitions, BellSouth could find that it is unable to participate in markets served by its more wide-ranging competitors, or that its "independence by choice" strategy is threatened by a hostile takeover. Investors took a critical view of AT&T's content-free strategy as soon as AOL and Time Warner announced their coupling. Mergers raise the bar for scale and scope.

Telecommunications service providers (and indeed the regulators that oversee them) recognize that the baselines for performance, scale, scope, and marketing will continue to rise as competitors exceed their own previous standards. The service provider that opts out of the escalation might not only fail to improve its market position; also, its market position can degrade considerably in comparison to competitors that continue to participate. Regulators appear to understand this trend as well. The mergers under review in 2000 would undoubtedly have been incomprehensible only 10 years earlier. Economic and industry movements have made them feasible in the global telecommunications market.

11.7 Supporting tools

Tools that can hone competitive parity skills often include competitive simulations or are sometimes outright games of strategy. Game theory and games in general often require the participant to think about the next move and then consider several moves downstream. Computer-based tools are well suited to assist individuals in developing these skills.

Decision support for competitive actions

Commercially available game theory software enables users to iterate through various scenarios. Software tools can enable the user to set parameters to capture the uncertainty of payoffs by deciding on the level of gain or loss associated with their choices. The software calculates the dominant strategy if there is one and the player can eliminate choices until the matrix produces the desired outcome. Game theory has introduced the notion that conflicts need not result in zero-sum outcomes. The best strategy for each player in the prisoner's dilemma does not result in the worst outcome for one or the other. Similarly, game theory simulations can identify cooperative rather than competitive solutions in situations of competitive conflict.

Other games in popular use can assist managers in developing skills to manage competitive activities. Both the Western game of chess and the Asian game of "go" have been used not only for business but also by military leaders for their war strategies. Another decision support technique is Monte Carlo simulation, which generates random values for uncertain variables and creates a model through iteration. Decision trees are also used to quantify the outcome of a series of decisions made between a service provider and its competitors.

11.8 Role of information technology

The primary use of technology to improve strategic decisions is the abundance of decision support tools available for management. Personal computing power has put very sophisticated tools and techniques on the desktop. Simple and flexible tools such as spreadsheets can assist service providers in calculating what-if scenarios. Game theory software standardizes the analysis process into a few parameters. Statistical tools can assist the planner in assessing the probability of various market responses to strategic initiatives.

A second consideration for information technology is in its structure. Successful market strategies have often been possible only because the service provider brought proper administrative support to the sustaining business processes. In many cases, this will require adaptable billing systems to meet competitive price structure changes and to introduce individualized packages that competitors cannot match. Furthermore, ongoing competitive analysis requires vigilance and prompt reactions, and this will require excellent in-house communications and access to competitive information from many sources.

Reference

[1] Beardsley, S. C., and A. L. Evans, "Who Will Connect You?" *The McKinsey Quarterly,* 1998, No. 4, pp. 18–31.

12

Distribution Strategies

We have seen some fascinating business models in Internet commerce—and "fascinating" is not always a complimentary term.

—Internet analyst

Distribution channels comprise the support structure required to deliver telecommunications services to customers. Elements in the distribution chain include wholesalers and retailers and may include any or all of the following additional participants: brokers, agents, resellers, and a host of other intermediaries. Telecommunications service providers can establish their market position wherever they want along the distribution chain and with whatever breadth they believe will make them successful. As with most other elements of the industry, the transformation from monopoly to a competitive environment requires a thorough reassessment of channel strategies. New entrants to the industry need to select the channels that leverage their strengths and maintain the discipline to focus within the distribution chain.

12.1 Differentiation within the distribution chain

The telecommunications services market has begun to differentiate, and one of the most visible points of differentiation is in the distribution channel. The wholesale channel is emerging as a source of growth in the industry. Phillips Group-InfoTech projected that the U.S. wholesale network service market will grow at an average annual rate of 24%, from $39 billion in 1998 to $116 billion in revenue in 2003. The market researcher anticipates that growth in wireless wholesale services, new entrants to the industry, and network service providers (including IP telephony and applications hosting services) will drive much of the additional revenue. Atlantic-ACM estimates that the U.S. long-distance wholesale market alone was $12.8 billion and shows double-digit growth [1], the most competitive long-distance market segment in terms of market share. Note that while the traditional wholesale channel is growing, much of the increase is due to new technology-based business models, such as ASPs and other supporting functions that did not exist even a few years ago.

The differentiation within the distribution chain is enhanced by the rise in service providers without a facilities investment, both interexchange resellers and CLECs. Wholesalers were largely unnecessary in the vertically integrated monopoly environment. They are not only necessary but also specialized and capable in the competitive market. Carrier's carriers such as Williams Communications in the interexchange market and MFN in the local access market recognized this opportunity. Both of these service providers have made a commitment to serving the wholesale market. Other utility companies took their lead from Williams (which was originally a pipeline utility) and leveraged their own rights of way and extensive conduit to create resellable local networks. Conectiv, a subsidiary of utilities Delmarva Power and Light and Atlantic Energy, sells capacity to IXCs in the wholesale channel and plans to enter the retail channel in the future. A Yankee Group survey stated that 65% of energy companies plan to develop their telecommunications assets into a separate business [2].

Pathnet specializes in second- and third-tier markets for its wholesale telecommunications solutions. Interexchange providers such as IXC Communications, Qwest, and Global Crossing set out to become wholesale providers but apparently changed their business models when they each acquired incumbent local service providers (Cincinnati Bell, US West, and Frontier, respectively). Canada's Teleglobe was originally a wholesale provider but moved into retail with its acquisition of U.S. telecommunications marketer Excel Communications. Now it is part of vertically integrated BCE (Bell Canada).

New business models in wholesale markets will boost the wholesale segment. Data CLECs work through ISPs and CLECs to offer broadband services to customers. The ISP or CLEC rather than the data CLEC most often brands the DSL services. ASPs, Web hosting services, and other network management segments often operate in the wholesale channel. Eventually, should the ASP market grow as anticipated, it will undoubtedly split into wholesale and retail segments.

Some ISPs have discovered that they can put their network investment dollars to better use as the competitive market has intensified. The rise of unbranded Web hosting services, network management, and messaging services has created a valuable supplier to the ISP industry. Besides reducing their capital investments for equipment and occasional upgrades, using a wholesale supplier enables the ISP to offer value-added services that might be unaffordable to support independently. ISPs can demand service-level guarantees and eliminate the staff and equipment that would be necessary to maintain these features in-house to gain the same service quality. Wholesalers include AT&T, GTE Internetworking, PSINet, and UUNet Technologies. As portals assume a larger profile in business and consumer markets, private-label ISPs leading to specialized portals could proliferate.

12.2 Coexisting wholesale and retail strategies

Sharing distribution responsibilities is relatively new to former monopolists, and local service providers have generally ceded these activities reluctantly. In a monopoly environment, there are many reasons to maintain vertical integration, which provides control of the entire distribution chain. At the outset of deregulation in local U.S. markets, most RBOCs remain in wholesale and retail channels with few announced plans to restructure their organizations permanently. Two local service providers, Frontier Communications and SNET, announced that they planned to separate their organizations into distinct retail and wholesale units. Soon afterwards, SBC acquired SNET and Global Crossing acquired Frontier. Since then, no restructures have occurred. Before the Frontier acquisition, Global Crossing had characterized itself as a "carrier's carrier," in other words, a wholesaler. Frontier, formerly Rochester Telephone, is undoubtedly in the retail business, although wholesale represented 30% of its overall business at the time of the acquisition. Either the acquisition represented a change in the business model for Global Crossing, or it represented a prelude to the divestiture of Frontier's retail business. SNET, on the other hand, is such a small component of SBC's growing portfolio that its previous restructuring plans have probably been shelved. British Telecom's announcement to restructure holds the potential to rejuvenate the topic for other major telecommunications service providers.

In the interexchange market, many of the largest carriers wholesale their network services, some more aggressively than others. Resellers of interexchange services expressed concern about the proposed MCI WorldCom and Sprint merger, fearing that the reduction of one major supplier in their market would create an obstacle to competition. Nonetheless, other carriers arose to fill the void of suppliers. AT&T, Sprint, WorldCom, Frontier, GST, and Qwest are aggressive wholesalers. MCI was never an active wholesaler, and WorldCom continues to take that responsibility in the combined company. The biggest concern for IXCs (and eventually for local service providers) is that providing wholesale services has the potential to cannibalize their retail base of customers. The decision rests upon a rhetorical question: whether wholesale recovers business that would have been lost to competitors anyway. For most service providers in the interexchange market, the wholesale operation is such a small part of the business that channel conflict is not an issue. If it were, resellers would still have an abundance of wholesale-only suppliers with considerable overcapacity.

Several telecommunications service providers plan to offer wholesale VoIP service. GTE Internetworking, Level 3, and Bell Atlantic have announced plans to compete in this emerging market, as well as Net2Phone, which is already in the retail IP telephony business.

Most of the RBOCs have inaugurated divisions to serve the carrier market, but no plans are apparent to uncouple the wholesale business further. To demonstrate a commitment to a new channel model, RBOCs could structurally separate the divisions or allow their retail divisions to purchase wholesale services outside (or build their own facilities), in preparation for an eventual divestiture.

Local telecommunications service providers that continue to serve both wholesale and retail channels in their core markets will face a variety of challenges. First, the situation creates channel conflict. Telecommunications service providers acting as wholesalers compete against their own customers when they operate simultaneously in the retail channel. Service providers further downstream on the distribution channel will always choose wholesalers that are not also competitors. AT&T recognized that its role as strategic competitor to local service providers was an obstacle to sales in its former equipment division. Channel conflict was one reason that AT&T gave up vertical integration in favor of corporate value when it divested its equipment manufacturer Lucent. Besides eliminating channel conflict for the equipment manufacturer, the divestiture of Lucent enhanced the flexibility of service provider AT&T to shop for technology from a variety of suppliers, an advantage already held by most of its competitors. Customers of suppliers in channel conflict often have a real or perceived impression that their supplier would like them to fail or that the supplier's priorities are directed toward its own retail customers.

There are many possible reasons that incumbent service providers remain vertically integrated as they move into deregulation:

- *The RBOCs do not want to reorganize until their major acquisitions and mergers are completed and integrated.* Both the scale and the pace of telecommunications industry mergers have strengthened. Merger integration is time-consuming but necessary and probably commands a priority over strategy-driven reorganization initiatives. Until service providers are more stable, there is no compelling reason to make a commitment to such a significant organizational change. In fact, if some mergers are borne more of opportunity than strategy, service providers might not know what their ultimate industry position is destined to be. Moreover, a service provider considering or planning to split its business into wholesale and retail functions might want to wait until the competitive local industry is more mature before making permanent commitments.

- *The incumbent local carriers believe that ownership of local facilities is necessary to maintain control over the quality of service.* Because incumbent providers already own local facilities, the decision to divest them carries different considerations than the decision that an emerging entrant would make about its entry into a new market. Owning facilities provides complete control over quality of service for calls originating and terminating within a service provider's network. Even some new entrants to local service have reconsidered their entry strategies and concluded that they need to own local facilities. AT&T expected to resell services from incumbents until it generated the critical mass it required to build its own facilities. After a flirtation with resale in the United States, AT&T changed its entry strategy and acquired local facilities in the form of cable properties. Whether its resale failure was due to the entry strategy or the specific relationship with the particular incumbent cannot be determined. In any case, AT&T apparently decided that it needed to control its own infrastructure. In the future, when wholesale providers are available and highly competitive, a similar level of control will undoubtedly be possible from network management software and service-level agreements. Even then, some service providers will choose vertical integration.

- *The RBOCs are in wholesale markets by decree and are not interested in creating a permanent wholesale line of business.* Becoming a vertically integrated provider for the long term is a viable strategy, especially if the business model does not include selling facilities at wholesale. Incumbent local providers are required to provide services at wholesale and could exit the business as soon as regulators allow them to do so. Still, the public statements of local service providers imply or state outright that wholesale is an important business for these service providers for the long term. If that is true, local service providers will undoubtedly revisit either their channel strategies or their organizational structures when they face competition in wholesale markets.
- *The RBOCs believe that they can run successful wholesale and retail businesses concurrently.* Local service providers that are successful in both markets will remain vertically integrated. Nonetheless, management will occasionally revisit the question of whether two successful divisions can be even more successful as independent providers. Furthermore, in a mature competitive marketplace, the wholesale efforts of a service provider whose core competencies are in wholesale services will undoubtedly prevail over those of the vertically integrated, extended service provider.
- *The RBOCs want to continue facilities-based operations indefinitely, anticipating giving up the wholesale business when competitive suppliers are abundant.* There are advantages to vertical integration if the service provider can manage or eliminate channel conflict. Telecommunications service providers need to look at their existing networks and decide whether it is better to upgrade their existing

facilities or start again with new technology. If their existing facilities would require less investment to make them competitive than to build new facilities when the wholesale requirement is lifted, the decision to wholesale for the short term is sound.

12.3 The case against vertical integration

Service providers that were not incumbents are therefore not required to provide local facilities to new entrants. For these providers, there are potential advantages to vertical integration, if the service provider does not provide wholesale services outside its own use. There are potential cost advantages, although this is not a guarantee. The facilities-based provider controls its own facilities and service quality, so the service provider seeking to differentiate its brand based on service quality will be tempted to build its own network. Many providers already own state-of-the-art networks, some of which have excess capacity.

On the other hand, there are several arguments against remaining vertically integrated while maintaining a presence in wholesale and retail channels:

- *If facilities-based incumbents expect the wholesale business to be successful, they should eliminate the channel conflict of a retail operation.* RBOCs are eager to dissolve the remaining restrictions on their businesses. The 14-point checklist preventing them from entering long-distance markets in their own territories would be nearly irrelevant if they separated their wholesale and retail divisions. Years ago, one competitive long-distance provider told regulators that it would support the RBOC entry into their business as soon as they split up their wholesale and retail operations. The strategic analysis of the viability of an RBOC—or for that matter, any service provider—divesting its conflicting businesses would include valuing the total of a focused wholesale division plus a retail customer base versus their current configuration. Williams Communications is the only remaining interexchange services provider in the United States that consistently states its commitment to a wholesale strategy. The company maintains that six of the largest 10 CLECs are among its wholesale customers and that the long-distance entry of the RBOCs will create additional business for the company [3].

- *A telecommunications service provider in highly competitive markets will find better uses for capital and expense dollars than in owning and operating a network in a market characterized by overcapacity.* Though it probably will not be true in a decade, customers largely view today's telecommunications network services as a commodity. This is less true in business markets than consumer markets and undoubtedly will change rapidly as packet-switched broadband IP networks replace today's circuit-switched ones. For the near future, telecommunications service providers are more likely to generate new customers and profits through their investments in differentiating their offerings through marketing and support rather than through facilities ownership. This issue will be compounded if local networks reach overcapacity, as they undoubtedly will do in the most lucrative urban and business markets. Telecommunications service providers that purchase their bandwidth wisely from wholesalers can potentially keep their costs lower than can those that commit to technologies and amortize their cost for years.
- *A service provider in the retail channel will be more flexible to enter and exit markets than a facilities-based competitor.* The resale price for local service is currently based on a specified discount against the incumbent provider's cost. Variations occur if the reseller chooses the entire facility or selects unbundled network elements. Still, resellers need to manage any nonnetwork cost to ensure that its cost profile is competitive or below that of the wholesaler in areas such as administration, marketing, and customer service. Therefore, in the resale model, costs are largely variable, while the facilities-based provider's costs are generally fixed. Resellers with insufficient market presence can exit relatively easily from any market. Facilities-based providers will enjoy a cost advantage once they build a base of customers, but they need to invest significant capital simply to enter the market. Facilities-based providers thus will undoubtedly remain in unprofitable markets longer than their more flexible reseller counterparts will.
- *A service provider without facilities investment can purchase services based upon newer technologies as they are launched without stranding investment.* While the facilities-based provider is waiting for its critical mass, it is underutilizing facilities that will eventually be obsolete. Building facilities takes time; even if there is a service backlog in a

market, investments will always precede revenues. The depreciation schedule for emerging network equipment is quite short compared to the copper-wire network. A service provider that decides to build facilities needs to account for the possibility that if the market does not materialize, or if a significant technological advance occurs, its equipment could be stranded well before it recovers its investment. On the other hand, a reseller (with favorable supplier contract terms) can switch easily to an alternative provider with newer technologies when they are deployed.

- *A service provider concentrated in a well-defined channel has a better opportunity to be excellent than one spread across the distribution chain.* Success in a highly competitive environment requires excellence, and excellence requires focus. Already, wholesale providers have made inroads toward differentiating their offerings in the wholesale market. Prior to its acquisition by Global Crossing, Frontier had made a significant commitment to its wholesale division with on-line provisioning and network reporting. Trouble ticket management, traffic analysis reporting, order status, and call detail records are available to customers, and bandwidth on demand was in the planning stages. Wholesale suppliers will enjoy economies of scale for operations support systems. They could potentially introduce features that would not be cost-effective for vertically integrated service providers whose emphasis is on satisfying their customers before improving the convenience of their own employees.

- *A service provider intending to grow through mergers will command more value for its customer base than its facilities in areas where the merger partner already owns facilities.* While carriers committed to remaining in the business serve the majority of the market, entrepreneurs often build startups with the intention of creating value and selling the business to a larger partner. For those new service providers, there is the potential that any investment they make in facilities will not be recovered by the acquirer's purchase price. The likelihood is high that an acquiring company already has an interest—and a facilities base—in the market that it is pursuing. Facilities-based acquisition candidates are liable to duplicate at least some of the facilities of the acquirer, which would have no value in the purchase price. Entrepreneurs seeking a short-term growth opportunity will avoid facilities investments except those that differentiate their services from all competitors.

Surprisingly, a 1999 study by Nortel Networks revealed that the valuation of publicly held ISPs without network ownership is higher than their facilities-based counterparts [4]. While it would be inappropriate to conclude that the facilities-based strategy is categorically invalid, it is certainly fair to deduce that facilities ownership is not required for ISP success.

12.4 Supplier and partner strategies

Other distribution strategies can include relationships with suppliers. Major vendors such as Nortel, Cisco Systems, and Lucent Technologies are financing new telecommunications service providers at rates that are well below the market rates of loans for new companies. The vendors benefit because their agreements with the borrowers often require large purchases of their own network equipment. They also undoubtedly assume that, as the service provider grows, inertia if not loyalty will steer their growing networks to additional equipment purchases well after the recovery of the loans.

Data CLECs that upgrade conventional copper facilities to broadband DSL have captured the interest and investment of telecommunications service providers. Investors in these new companies include Verizon,

TABLE 12.1
Values of ISPs Without Networks Are Higher Than Network-Based ISPs

	ISP	Valuation (Millions of $)	Subscribers (x 1,000)	$/Subscriber
Own network	Erol's	$83.5	293	$285
	UltraNet	$27.0	32	$844
	Netcom	$321.0	540	$594
Purchase from wholesaler	Earthlink	$593.4	400	$1,484
	Mindspring	$440.6	280	$1,574
	AOL	$12,500	11,000	$1,136

AT&T, MCI WorldCom, Qwest, and Frontier. In addition, many data CLECs are integrating their order processing systems seamlessly with the local service providers to which they interconnect. This initiative reduces the time it takes to provision services, eliminates manual rekeying of data, and improves the documentation for the transaction. Third-party software developers have recognized the potential of this market as the industry becomes more specialized. The result will be that even the smallest service providers will probably have access to sophisticated electronic transactions with their business partners.

Telecom Finland leveraged its in-house technical expertise and the enthusiasm of its supplier to launch the world's first ATM service in 1994. Cisco Systems sent the service provider early releases of its equipment in exchange for feedback and other support, benefiting both partners. The manufacturer receives valuable market response to its systems, and the telecommunications services provider establishes a first-mover advantage in its own market on its own terms. This form of partnership is likely to occur more frequently when the market develops more fully and differentiated services emerge.

In Europe, telecommunications service providers entering new markets sometimes collaborate with local companies to ease their entry. Partnerships with rail companies can offer access to rights of way. Other European service providers have established relationships with local retailers or utility companies to provide access to distribution channels.

Supplier partnerships will become more important as companies expand their outsourcing of mission-critical functions. Suppliers will need to provide real-time operations data, services-on-demand, and just-in-time procurement. Advances in information systems and the rise in suppliers to meet market requirements will foster the growth in outsourced services and strategic partnerships along the distribution channel.

12.5 Internet strategies

Distribution on the Internet has captured the attention of most companies, and telecommunications providers are no exception. The largest telecommunications service providers, surprisingly, have not been at the forefront of e-commerce for their own businesses, although they do offer Web hosting and other services for their customers. While the telecommunications industry spends more on information technology than most other industries, much of the expenditure until recently was directed toward nonstrategic areas such as legacy system maintenance. An *Information Week* survey determined that the telecommunications industry is average in nearly all categories—new technology spending, e-business spending and sales, data capture, and electronic supply chains. The industry has above average performance in other strategic categories. The telecommunications industry is above average in spending on customer relationship management and Web-based customer support. The telecommunications industry is in first place in selling products and selling services on the Web.

Web users constitute a small but growing group of customers, but they generally comprise the largest users and the most profitable customers. Whether telecommunications service providers delayed their Internet strategies to meet the technology adoption curve or simply had other priorities is no longer relevant. Service providers are actively compensating for lost time with on-line strategies.

Telecommunications service providers can reduce their labor costs and improve the availability of information to customers by moving routine transactions to the Internet. More than a decade ago, France Télécom introduced the Web-like interactive information service Minitel to its customers for information and transactions. Today, most U.S. local and long-distance service providers offer on-line review or payment. On-line Yellow Pages services offered by RBOCs were once the mainstay of their e-commerce efforts long after other Internet-based providers offered e-commerce. Canada's New Brunswick Telephone used directory services to transform its application from a simple automation of paper products. The service provider encouraged users to find listings on-line by pricing the service aggressively, offering more information than would be available on paper, and updating the service twice daily. Furthermore, the service provider enables customers to view and pay bills on-line, change the settings of services, and order new services on-line. Frontier offers similar on-line capabilities to its medium-sized business customers. BellSouth states that 12–14% of its midsize business customers use the on-line channel, although on-line services have attracted only about 5% of consumers. Bell Atlantic enables customers to subscribe to ISDN and other enhanced services on-line, and GTE uses its on-line presence to sell products such as telephones and caller ID boxes.

SBC Internet Services uses an on-line help application to enable users to search for an answer to a technical support question. If the user's question is not resolved through the search engine's exploration of the knowledge base, the customer can chat with a live representative through text messaging. Audio and video transactions are the next step in on-line customer support.

Bell Atlantic and others also permit customers to determine if their distance from an equipped central office is within the acceptable range for DSL service but stop short of on-line provisioning. Bell Atlantic's first bundled offering upon its entry into the long-distance market included an Internet-administered service. The low prices for the service were only available to customers willing to forego a paper bill. Frontier uses a reduced per-minute long-distance rate to encourage customers to order service on the Internet. MCI WorldCom has offered a 2% discount to small business customers who pay their bills on-line within the first five days of the billing cycle, and a 1% discount within 10 days.

Frontier, Sprint, and Concentric Networks Corporation provide businesses with on-line information beyond conventional customer care. Service providers find that on-line services represent an opportunity to differentiate their offerings to the highly contested large enterprise market. Covad's on-line provisioning system enables customers to order or upgrade DSL services through a Web interface directly into its automated provisioning system, while other DSL providers offer similar features or plan to do so. Interpath, a CLEC serving the business segment, enables its customers to change their service parameters through a Web-based portal.

ASPs represent a new form of on-line distribution, and telecommunications providers in several segments—ISPs, local service providers, and CLECs—are exploring opportunities and partnerships in this potentially huge market. The breakthrough in the ASP market (not restricted to telecommunications service providers as suppliers) is that the product itself, not simply its administration, is Web-based.

Channel conflict threatens service providers that allow their customers to bypass the conventional intermediaries such as retailers, resellers, and agents. The customer with access to preferable prices, service levels, or terms of service will seek the service provider that resides the farthest back in the distribution chain. Wholesalers would be able to sell directly to customers without the participation of their distributors, a short-term gain with a significant long-term weakness. Manufacturers often eliminate this problem by providing information about dealers and purchase locations on their Web sites without offering on-line shopping. As the technology matures, telecommunications service wholesalers or others that do not normally sell directly to customers could develop on-line links to their distributors for on-line provisioning. These links, utilizing call centers or other customer service organizations, can make the origin of the transaction transparent to the customer.

12.6 Supporting tools

Because distribution channel decisions often require choices between channels, it is imperative that decisions have the support that tools can provide. While in the past small firms might have based these choices on instinct, modern managers can collect detailed data and draw more reliable conclusions.

Vertical integration framework

Strategy consulting firm McKinsey presented several industry characteristics that encourage vertical integration [5]. These include the following:

- The distribution chain is risky, mostly because of too few buyers or sellers.
- An imbalance of market power exists between buyers and sellers.
- There is an opportunity to create market power through barriers to entry or price discrimination.
- The market lacks required stages in the distribution chain.

While the overall telecommunications industry lacks most of the foundation for a sound vertical integration, some segments of the industry are undoubtedly better candidates for this strategy. The strategy consultancy proposes a framework of decisions facing the innovator of new technology and the conditions favoring vertical integration. Another decision tree provides direction for establishing varying degrees of vertical integration. Finally, the decision process identifies the particular analyses required before making the vertical integration decision.

Data mining

Data mining (or data warehousing) is a popular tool used by telecommunications service providers to uncover information for many purposes, including distribution channel strategies. In the United Kingdom, for example, the main source of wireless churn is in the distribution channel [6]. One wireless service provider discovered that its own reward structure was causing churn.

For an application directed toward improving channel management, the service provider populates the warehouse with relevant information and then interrogates the data to determine the success of distribution channels in use, or channels in combination. In its simplest form, the data warehouse can identify the distribution channel that generates the highest rate of churn in its customer group. More sophisticated analysis can rate the performance of various channels in terms of metrics such as acquisition cost, lifetime customer profitability, fraud, or uncollectibles. The warehouse can generate reliable assessments of the productivity of a specific agency or a specific representative as compared to peers. Furthermore, the ability to add criteria to an inquiry can bolster this evaluation by controlling for external factors such as the quality of a sales territory or the service packages available to the agent.

12.7 Role of information technology

Information technology is essential to modern distribution strategies. Extranets are required for communication with both suppliers and customers. Information and communications technologies enable telecommunications service providers to procure equipment with just-in-time delivery and manage their transactions with suppliers, intermediaries, and customers. Internet distribution will replace all other customer interactions for many customers. Price-sensitive consumers and small business will appreciate their ability to configure their own services on their own schedules. Business portals will enable enterprise customers to manage their own networks through information reporting and on-line reconfiguration. Ironically, the use of the Internet channel will also facilitate outsourcing of many functions. After all, if a transaction is electronic and therefore not physically present in either space or time, it is not relevant who operates the transaction, as long as the transaction meets standards for responsiveness and accuracy. These are just some of the possible distribution opportunities made available with technology. One advantage of investment in transforming the distribution channel is that most expenditures result in reduced costs, either from reductions in labor or improvements in the quality of information. Improving the quality of customer service, always a difficult metric to quantify, is probably not a necessary component to justify these initiatives.

Inside the telecommunications service provider, channel strategies can also be improved through technology. Data mining consumes state-of-the-art processing power and requires responsive networks connecting users and enormous storage capabilities when systems are very sophisticated. On the other hand, less sophisticated systems will be available that can deliver slightly less processing power for considerably less investment. Service providers will need to fit the data mining function into their overall strategies and decide whether world-class performance in this area is necessary for success.

References

[1] Smith, J. R., and W. Park, "Sizing the U.S. Long-Distance Market," *Phone+*, Vol. 13, No. 14, pp. 68, 70–72.

[2] Henderson, K., "Resale Energizes Power Company Entry Into Retail Telecom Markets," *Phone+*, Vol. 12, No. 3.

[3] Masud, S., "Williams: Second Time a Charm?" *Telecommunications*, Vol. 33, No. 2, p. 22.

[4] Deng, S., "Why Build, When You Can Outsource?" *America's Network*, Vol. 103, No. 2, p. 78.

[5] Stuckey, J., and D. White, "When and When *Not* To Vertically Integrate," *The McKinsey Quarterly*, 1993, No. 3, pp. 3–27.

[6] Siber, R., "Combating the Churn Phenomenon," *Telecommunications, International Edition*, Vol. 31, No. 10.

13
Pricing

Why does it cost more to fly from Idaho to Oklahoma than it does to fly from New York City to Florida?

—RBOC sales manager

For the first century of monopoly telecommunications service, pricing followed the principle of "universal service," which simply eliminated through subsidies the large entry costs for consumers to enjoy telecommunications services. Business subsidized residential service; long-distance subsidized local services, and densely populated areas subsidized rural areas. Any service that enhanced basic communications provided high profit that could contribute to the subsidies. Deregulation adds complexity to the equation. Competitors, unlike monopolists, do not control prices; markets do. Complicating the situation, Internet telephony, by design, does not contribute subsidies, as do conventional long-distance services, yet this new technology is finally a viable substitute service. This creates an arbitrage opportunity and threatens to erode the remaining subsidies in the network. In the most competitive consumer markets, including long-distance, wireless, and Internet access, revolutionary pricing redistributes market share. Innovative pricing structures that reflect customer desires, meet universal service objectives, and maintain profitability will be a key differentiator in the telecommunications market.

13.1 The pioneers of pricing

It took about 15 or 20 years of long-distance competition for any of the service providers to challenge the Byzantine time and distance-sensitive pricing mechanism. Until Sprint introduced a per-minute rate anywhere in the United States in the mid-1990s, customers placing a long-distance call simply did not know what the call would cost. Most calls depended on mileage bands that did not correspond to any other familiar measure. Operator services, most of which did not require a live operator's intervention, such as the customer entering a calling card number on the keypad, represented an additional charge. While today's low rates diminish the impact of pricing uncertainty, at that time domestic calls within the United States cost several dollars at a minimum.

Sprint introduced per-minute pricing, and the proof of its value is that competitors immediately responded with per-minute prices of their own. Indeed, Sprint is often the leader in pricing innovations. In 1998, Sprint introduced a plan it called Sprint Unlimited. For a flat fee per month, customers have unlimited calling throughout the weekend. The plan targeted very high-volume users. After all, competitive prices were a low per-minute rate on weekends from the other major IXCs. At the time of the service launch, the customer needed to consume more than six hours per month of weekend calling simply to break even if the competing plan holds an administrative charge (as most do). Weekends that the customer does not originate calls from home create pressure to use even more minutes during the remainder of the month. Furthermore, customers are often unaware of the exact hours in which the discounts apply. Instead, they use the long-distance service freely, and daytime rates can be five times the discounted charges.

Once customers understood what they were paying, price competition became more viable. After decades of competitive distance-sensitive pricing, it took only a few years to drop the price of a call by up to 75% of Sprint's original per-minute price. The simplicity of the new pricing made it possible for customers to compare service providers. The competitive intensity of the market then forced long-distance providers to design competitive service plans.

The ghosts of monopoly pricing haunt all service providers, especially incumbents. Unlimited local service does not recover its costs in most consumer markets. Complicating the problem is the rising use of the Internet and the unlimited monthly charges of ISPs. Customers who might have been receptive to measured pricing in the past are much more resistant, now that they or their families spend hours on-line. In most European countries, local service was measured, and that pricing arrangement is often cited as the reason that the Internet had flourished in the United States and grown more slowly elsewhere. One German provider, Mobilcom AG, devised a flat-rate local service with several conditions (such as time-of-day and a restriction that the called party also be a Mobilcom customer) unheard of in the United States. The interest in this service and its flat-rate Internet access demonstrates that pent-up demand for unlimited services exists, even as European customers are accustomed to usage-sensitive pricing.

AT&T's innovative Digital One Rate pricing structure not only catapulted the service ahead of its wireless competitors, but it also created a viable pricing alternative to landlines. A customer contemplating a second line for Internet access could use the existing landline instead and use wireless service for incidental calling, including long distance. According to the Yankee Group, wireless prices dropped by about 40% between 1996 and 1999. The "wireless premium," the ratio between wireless and wireline prices, is at only 2:1 for those consuming 1,000 minutes per month and at 3:1 for 500 minutes. In countries such as Israel and Finland, lowering the wireless premium results in significant landline displacement. Denmark's low 2.06:1 ratio has encouraged a robust wireless industry and significant landline displacement. The Yankee Group concluded that wireless users in the United States already displace about 12% of their wireline traffic, an amount that will grow to 17% by 2001. This is significant to wireline providers. First, they stand to lose second line orders to wireless providers. Most wireline service providers place second lines to the home at the time they install the first lines. Thus, they will not recover their labor, nor will they capture the recurring revenue stream if a different wireless provider (mobile or fixed wireless) captures the customer's business. Second, if wireless providers continue to copy the AT&T pricing model, customers will discover that their overall costs will drop if they buy a large bundle of wireless minutes and migrate their long-distance calling to the wireless telephone. Long distance is the most profitable component of bundled service and is sometimes the only source of profit. Gaining the local customer without the attendant long-distance minutes can be less appealing than losing a customer outright.

13.2 Pricing and profitability

Like the airline industry and banking, perishable services, high fixed costs, and high exit barriers characterize the facilities-based telecommunications industry. Services are perishable because any time a network is available but not in use, the opportunity to recover costs is gone. Network construction and maintenance costs have dropped but will still represent a significant cost element for facilities-based providers for a long time, especially because the other large cost, labor, is also dropping fast. High exit barriers exist because service providers need years to recover their network investment. Leaving underperforming markets is not an easy decision.

The result of these attributes is that telecommunications service providers need to sell their capacity aggressively, at prices that will recover costs. There are challenging tradeoffs here, because the commodity nature of telecommunications services pressures service providers to sell at the lowest prices possible. One implication is that service providers need to find ways to sell unused capacity. While steep discounts are always an option to dispose of excess minutes, somehow the service provider needs to accomplish this without jeopardizing the nondiscounted revenues from less price-sensitive customers.

Another challenge in the information industry, and indeed the telecommunications industry, is that incremental costs approach zero for many services. The high fixed costs of long-haul networks create an almost zero cost for services sold above the break-even point. Many central office or server-based enhanced services have few costs beyond their development costs. On-line publications have no distribution costs for the incremental user and in fact increase in value as the user base expands. While these characteristics do not conform to traditional economies of scale, they imply that successful facilities-based providers need a critical mass of customers to maximize their profitability.

Regulators and service providers in the United States have worked to reduce long-distance access charges and pass along the savings to customers. In 1992, access charges were 5.8 cents per minute and by 2000 dropped to 3.3 cents. Access charges could drop to one-third of that level. Fierce industry competition tends to pass any cost reductions to customers, and access charges do account for much of the decrease in long-distance costs. Furthermore, some new carriers make the persuasive argument that access charge subsidies based on historical costs overstate prices and actually hinder competitive new technologies, such as fixed wireless.

The falling costs of underlying technologies (partly due to their increased capacities) are also responsible for price reductions. So far, long-distance services have proven to be quite price elastic, that is, falling prices create more calling volume. Service providers that can reduce their prices without reducing their margins will increase profits. On the other hand, the competition is so intense that service providers are forced to shave their margins, undoubtedly hoping to maintain adequate profitability with the lowest prices possible.

Traditions are a strong foundation for pricing structures, and some traditional pricing mechanisms for voice services do not match costs. For example, the cost of the last-mile local facility is a large one-time investment, historically recovered through monthly rates. The mobility of customers (and the imminent switching from one provider to another in a locally competitive environment) will drastically reduce the average duration of the life cycle of the local customer. This will prevent service providers using old algorithms from recovering their facilities investments. Coupled with drastic reductions in technology costs, it might soon be possible for service providers to charge a flat one-time fee for installation and a minimal price for local service in densely populated areas. Similarly, local service providers, inundated with long Internet sessions on an unlimited local line, are seeking ways to reroute the traffic from usage-sensitive ports to fixed-cost facilities.

On the other hand, new entrants are creating new traditions. ISPs changed customer expectations when they moved toward unlimited pricing structures. Pioneers in voice communications pricing, generally underdogs, offer unlimited packages. Unlimited IP-based long-distance minutes are available through USATalks.com and Cybertel. Australian service provider Telstra recognized that its customers were reluctant to make international calls, no matter what the price. It introduced flat-rate unlimited calling to the United Kingdom (about $10 per call), which encouraged customers to increase their calling, satisfying customers with discounted service and giving the service provider better utilization of its international facilities. Broadmedia introduced a package of unlimited IP-based international calling for a flat rate per month. IP voice provider delta-three.com allows customers to place calls from an equipped PC anywhere in the world to U.S. and Canadian telephones for a small flat rate per month. The package also includes hundreds of minutes of PC-to-phone IP calls to selected cities in Europe and Asia. Flat-rate packages are not limited to emerging telecommunications providers. Both AT&T and Sprint offer a flat-rate-per-month service in Canada that has been so popular it has reportedly exceeded the capacity of the network [1].

The impact of these offers will be groundbreaking. Customers, other than those that are extremely budget-conscious, have expressed great interest in flat-rate services. Technology issues prevent the quality of most of the unlimited IP-based packages from comparison to traditional networks, but innovative service providers will solve these problems soon. The ubiquity of the Internet means that no monopoly provider can be protected from favorable prices anywhere in the world. Unlimited service, or something very close to it, is the future of telecommunications.

Some wireless service packages are moving toward unlimited usage for their highest volume customers. Wireless North offers a flat rate per month unlimited wireless service specifically targeting business customers. A similarly equipped wireline customer pays close to the same rate per month to US West, the incumbent local service provider [2]. US West responded with competitive pricing to stem the flow of churners. Wireless providers AT&T, Sprint, Nextel, and Verizon already offer premium packages with so many bundled minutes as to be virtually unlimited for voice communications. Cricket Communications, a subsidiary of Leap Wireless, offers an unlimited access option without roaming for a flat rate per month.

For data services, wireless customers have expressed little desire to pay by the byte, as some service providers have proposed. A survey by Cahners In-Stat Group determined that 73% of business professionals want to pay a flat monthly rate for wireless data access. Only 8% of respondents preferred usage-based pricing, and a paltry 3% would like to pay by the packet. Wireless providers cannot rely on the assumption that competitors will be unwilling to offer unlimited packages. Wireless data reseller GoAmerica offers an unlimited cellular digital packet data (CDPD) wireless package for a flat monthly rate.

Cable companies moving into telephony will need to harmonize their traditions of unlimited pricing for television with a need to innovate against entrenched providers. Some will choose to bundle some long-distance services into their local calling packages. Certainly AT&T's TCI and MediaOne units will take their lead from the parent company's wireless innovations.

Another incentive to offer unlimited service is that it simply reduces costs. The cost of billing is 11% of long-distance revenues [3]. Eliminating the measurement passes on savings to customers, on the condition that access charges and intercompany settlements, if they remain, can be estimated in some other way. Customer service costs will decrease, as customers are less likely to dispute a monthly charge than a single forgotten or incorrectly billed call. In conjunction with on-line bill presentation and electronic payment, billing and collection costs are likely to recede until they are negligible.

Incumbent telecommunications providers are also taking the initiative to transform price structures by eliminating traditional but arbitrary service distinctions. In mid-1999, Telenor, Norway's incumbent provider, designated the entire country to be a single rate zone, thereby eliminating the distinction between local and long-distance services. MCI WorldCom's chairman announced an all-distance plan in the competitive New York market in which local and long-distance minutes are undifferentiated. Ameritech Illinois filed to equalize prices for business and residential services, a longtime source of price discrimination.

13.3 Competing against free service

The newest inducement in the information industry is the abundance of free services. Wireless service providers give away handsets and enhanced services in the hope of selling profitable minutes of use. Computer manufacturers give away Internet access and Internet access providers give away computers. Other entrepreneurs give away hardware, software, and Internet access simply to display their own free portals to targeted customers. How does the service provider survive selling in a market in which its basic offer to customers is free from competitors?

This challenge is not a trivial one for telecommunications providers. The unit cost of some fundamental service components, such as bandwidth, already approaches zero. The forces of competition, technology advancement, and competitive pressures will squeeze costs further. Customers are influenced more by price than any other present service differentiator. Furthermore, competition from free services comes in from everywhere.

Free Internet service is a real enabler in Europe, where customers pay per-minute rates for local calls. Measured local service changes adoption patterns for new technology. Looked at another way, Americans enjoy free—or at least unlimited—local service. That price structure probably created the conditions in which the Internet would thrive quickly in the United States. On the other hand, Europe's measured local service acted as an enabler of similarly measured (and priced) wireless service, creating high penetration compared to the United States. This demonstrates that while free or unlimited offers do create markets, they only create them as substitutes for existing markets and existing pricing structures.

Tiny Computers, Britain's largest domestic computer manufacturer, offered a free PC for consumers that signed up for a telecommunications service provided through the company's reselling arm. AOL made a similar offer to customers who were willing to take a three-year subscription. In the United States, PC manufacturer Dell takes the opposite approach, including a year's free Internet access with the purchase of a full-featured consumer PC. This offer is more of a promotion than a bona fide giveaway, and simply uses the AOL model of free introductions before asking for subscriptions. Companies like Dell could find it worthwhile to introduce new users to the Internet, perhaps to lessen the first-year load on their voice-based customer service group. These offers will represent a viable competitor for those companies working to build a base of Internet subscribers.

In the United Kingdom, Freeserve was the first ISP to offer free access, and it gained one million subscribers (or 40% of the U.K. market) within six months of its service launch [4]. Freeserve's motives were not completely altruistic. Until its parent company divested the ISP in 1999, the service was a division of Dixon's, the country's largest electronics and computer retailer. Other retailers, recognizing the distribution and selling opportunity, followed suit, including retailers Toys "R" Us and entertainment retailer Virgin. AltaVista, a U.S.-based portal, announced free Internet access and unmetered calls to U.K. customers for a one-time fee. Cable provider NTL immediately matched the AltaVista offer with no added charges; then other major ISPs followed, including industry pioneer Freeserve. In Spain, the number of Internet users nearly doubled within two months after local ISPs began to offer free access. Icero began free Internet service in Argentina in late 1999. The startup intends to build a revenue stream through advertising and eventually e-commerce. StarMedia Network, Inc. and other partners announced free Internet service in Brazil, Argentina, Mexico, and other Latin markets.

Freeserve and other pricing pioneers launched a revolution in Europe. In the United Kingdom, incumbent provider British Telecommunications soon announced bundled call and line rental packages that offered unlimited free calls to ISPs at defined times during the week. Incumbent providers in Spain, France, and Ireland have taken similar actions. Telecom Italia's Tin.it gave up subscription access in the face of free competition. Several Danish providers offer free Internet access. Swedish ISP Spray offers free service in non-English-speaking countries in Europe.

In the United States, unlimited local service launched the Internet market without the help of free ISP access, but entrepreneurial service providers are willing to make the investment in eliminating subscription fees to acquire market share. NetZero is the best-known advertiser-supported free ISP in the United States. The user sees a small display window on the computer screen delivering targeted advertising, based on a profile the user provides at sign-up. GTC Telecom promoted its own new Internet service by offering free access to up to 100,000 customers. Its ongoing plans include aggressive discounts for long-distance or free Internet service when bundled with its more conventionally priced long-distance service. Portal AltaVista offers subscription-free access in the United States. Already the tenth most-visited site on the Internet, AltaVista expected that free access into its portal would create more users for its shopping, information, and search services. Larger service providers are beginning to notice the attention paid to service providers that offer free services. Qwest offers free Internet access when a customer purchases long-distance service.

BroadPoint Communications and Duquesne Enterprises (an affiliate of Duquesne Power & Light) introduced FreeWay to provide free long-distance service. Customers receive two minutes of free calling in exchange for 10 or more seconds of advertising. Upon subscribing, customers complete a questionnaire detailing their demographic characteristics and interests. Two types of interactivity with advertisers are possible: The customer can press a button and speak with a sales representative or press a button and get an e-mail message about the product. The service provider reports response rates of 3% for service representatives and 10–25% for e-mail requests [5], rates that compare very favorably to alternative direct marketing response rates.

Sweden's GratisTel International offered free mobile phone calls in exchange for listening to advertising. One challenge facing these providers is that advertisers require targeted advertising in exchange for their subsidies. Another challenge is that countries such as Germany and Austria prohibit advertising-financed telecommunications services.

Another business model was launched by RhinoPoint.com. Its service simply pays the user's Internet charges in exchange for the customer's completion of a marketing survey once a month. Customers also pay RhinoPoint.com a one-time charge.

To protect its profitability when its competitors draw away customers with free offers, a telecommunications service provider can do the following:

- *Compete with unadorned free services against the competitor but provide a migration path to feature-rich services for pay.* AOL, once Freeserve had overtaken its first-place position, reversed its long-argued position and offered its own free Internet service. AOL's free service does not have all of the features or content of its subscription-based alternative. In fact, AOL's dual offerings represent a sound approach to competing in an information market. The technique of offering two versions of a service or product is quite common in the industry. Hardware manufacturers often sell two versions of a computer or chip by disabling functions in the lower-priced product. Software manufacturers enable customers to upgrade to a more capable version of the program with a software key that unlocks the hidden capabilities. Information providers such as Dow Jones offer a variety of access mechanisms to their archives and price them accordingly.

- *Concentrate on core competencies in areas that are not commodities and therefore less vulnerable.* If Internet access becomes such a commodity that competitors can absorb its costs as the price of acquiring customers, then it is a candidate strategy for any telecommunications provider. In addition, service packages can include commodity services in addition to differentiated services. Telecommunications service providers will be able to create full-featured bundles for a total price that covers all of the costs of the components. In fact, if nonservice-provider competitors can offer a service at no charge that represents a significant cost to a service provider offering it on its own, the service provider should recognize that its cost profile is unacceptable. Indeed, the free service offered is an excellent candidate for outsourcing.

- *Investigate alternative pricing structures that will recover costs and attract customers.* The upheaval in global telecommunications markets creates an excellent opportunity to revisit many assumptions buried in price structures. Notwithstanding universal service regulatory requirements, some fixed costs are still recovered through usage charges and vice versa. These pricing anomalies invite competitive arbitrage. Telecommunications service providers should transform their price structures before competitors take customers from the most lucrative segments. It will be more difficult to restructure prices when a larger proportion of customers benefit from pricing idiosyncrasies.

- *Exploit the growth in free services to promote the purchase of enhancements.* The growth in Internet access in Europe will force incumbent providers to restructure their local service pricing, but it also provides an unprecedented opportunity to sell broadband access. Service providers recognizing the expansion of the market need to leverage the opportunity by creating promotions and distribution channels to capture the rising interest and money newly unspent on free services.

- *Recognize which free services will probably fail if the price gap becomes insignificant.* Free long distance in exchange for the inconvenience of listening to advertising was a much more attractive service when per-minute charges were $0.25 than at current competitive rates. Free PCs that avoid a charge of what was once $1,000 or more are not as attractive when companies provide free Web-enabled systems to their employees for home use or when the price of systems drops to only a few hundred dollars. Free Internet access became available only a few weeks after the leading telecommunications service providers dropped their monthly subscription fees from about $30 to under $10. ISPs are working hard to rebalance the revenue stream so that advertising and purchase commissions cover their operating costs. It is possible that the Internet model will eventually follow the television model, which includes advertiser-supported broadcasts at no charge and premium uninterrupted programs for a charge. When competing against free services, as in commodity markets, it is critical to develop brands and otherwise differentiate one's own services to attract attention and revenues in a crowded marketplace.

13.4 Telecommunications as commodity

The proof that consumer telecommunications services are still a commodity is in the market research data concerning price. After the deregulation of the U.S. interexchange market, 91% of customers leaving AT&T named price as their reason for leaving [6]. A Booz-Allen survey of European customers found that 89% of business customers and 75% of consumers cited lower prices as a reason to switch service providers. Even for business customers, expected to require excellent service, only 49% of respondents cited service as a reason for switching providers.

Wireless customers churn in spite of overall customer satisfaction. More than half of all wireless customers stated that they are very satisfied, and 88% are at least generally satisfied [7]. Still, 70% would consider churning for a 10% price reduction and more than 60% would switch carriers for a free telephone upgrade.

Service providers are tempted to launch price wars to gain share. In the telecommunications industry, lower prices also increase usage, on networks that are commonly underutilized. The rationale, then, is that with or without competitive retaliation, price reductions that reduce margins are worthwhile. In practice, price wars tend to benefit only the customer. First, initiating price reductions launches the prisoner's dilemma matrix whose conclusion for wise competitors is to match the decrease. Therefore, a service provider that reduces prices without any internal incentive (such as capitalizing upon business process efficiencies) will likely face competitive reaction. Furthermore, externally driven price reductions, due to technology cost reductions and the phaseout of access charges, will already create incentives for all service providers to lower prices and result in higher usage. Moreover, increased Internet penetration and faster data services will increase backbone network utilization, if new bandwidth does not outpace the growth in consumption.

Price discrimination occurs when two customers pay the same price for products of different value. The wireless customer that turns off the bundled voice mail option in the service package does not receive a bill credit; price discrimination exists. Alternatively, price discrimination is present when customers pay different prices for identical products. The telecommuter paying a low residential rate for a home telephone and a high business rate for exactly equal services on a second line in the same room is experiencing price discrimination.

Monopolist telecommunications providers often built price discrimination into the business and consumer rates, time-of-day pricing, and data or fax lines. Competitors can also include price discrimination in their rate structures, but they face a significant difference from the pricing choices of monopolists. The customer experiencing monopoly prices has no competitive alternatives. Overpricing desired services in a monopoly environment is practical and useful. Overpricing services when alternatives exist creates opportunities for new entrants to serve only the most lucrative markets and benefit from the arbitrage opportunity.

MCI started its own business by serving only business customers with service only between the heavy Chicago to St. Louis long-distance route. AT&T could not respond competitively on the single route, as its rates were nationwide and distance-sensitive. In any case, regulators would undoubtedly have prohibited any competitive response by AT&T that favored a single competitive market. Ten years after divestiture, AT&T's prices for the high-volume business market and small business customers had fallen from 1984 levels due to intense competition. Nonetheless, the service provider was able to raise rates for the mid-sized business customer, a market that had not yet been a target of competitors.

Competitive telecommunications providers already use price discrimination to improve their profitability, and creative new entrants have already found opportunities to differentiate their offerings through pricing and price discrimination. Virtually every long-distance plan encourages customers to use the network outside of business hours. After Bell Atlantic received permission to offer long-distance services to its local customers, its first bundled offering in the highly competitive New York market included a package that offered the customer low rates in exchange for on-line account administration. In theory, the lack of a paper bill reduced the service provider's cost, and the on-line account center improved customer service for those customers who prefer to conduct business on the Internet. In application, the on-line center will not reduce Bell Atlantic's cost until a critical mass of customers is on-line. Bell Atlantic is encouraging consumers to move on-line through reduced prices on services.

All of the major long-distance providers have found the small monthly charge to be a valuable source of price discrimination. The charge, not applicable to any calls, reduces the appeal of discount programs to all but high-volume customers. In an environment of very low per-minute rates, a $5 charge would, for example, double the effective rate of a call for anyone averaging 100 long distance minutes per month on a $0.05 per minute plan. As late as 1998, the Yankee Group estimated that 75% of the population did not subscribe to a calling plan [8]. The reasons customers ignored the potential discounts included a perceived lack of real savings, the fact that plans were not important, or a fear that the customer would not qualify. Therefore, customer inertia is potentially the best inducement for price discrimination used by long-distance providers.

13.5 Revolutionizing price structures

AT&T has used bundling effectively as a differentiator and a churn reduction mechanism. AT&T offered large discounts to customers billing the call on its Universal credit card. The company also leveraged its investment in satellite television provider DirecTV by offering favorable terms of credit on the Universal credit card. Moreover, customers of long-distance services received a free pay-per-view movie each week they subscribed to AT&T's long-distance services. In the same promotion, customers earned a month of free DBS service for each year they subscribed to AT&T's long distance.

AT&T also bundled consumer long-distance service with its highly successful Digital One Rate wireless plan. Customers subscribing to Digital One Rate can receive AT&T's best long-distance rate on their home telephones without paying the minimum monthly charge. All of these mechanisms serve as switching costs. AT&T believes that each of these programs has helped reduce churn. There is an interesting component to this form of bundling. In each case, AT&T uses its differentiated service to sell its commodity service. For the pay-per-view promotion, AT&T created a switching cost to the customer (the retail price of the movies) at far less cost to itself (the wholesale cost, which approaches zero). In the free service promotion, AT&T earned year-by-year loyalty in a market characterized by 30–40% churn. The switching cost to the customer was enough to engender loyalty, especially when providers are virtually interchangeable other than price. Again, the cost to AT&T was insignificant compared to the potential cost of churn.

In the case of Digital One Rate, there were few competitors with equivalent services. AT&T differentiated its calling plan from competitive wireless services. AT&T could leverage its growing wireless customer base into additional market share for its commodity consumer long-distance service. AT&T offered its best long-distance rate, without monthly fees, to Digital One Rate customers. An added benefit for AT&T and its customers was that the company administered services on a single bill, enhancing convenience for customers and reducing AT&T's administrative costs. Another advantage of the Digital One Rate plan was its simplicity compared to its competitors. AT&T has been very successful competing against per-minute long-distance charges that are lower than its own offer, because customers seek simple, comprehensible services. The service provider that effectively bundles services and includes easy customer care will capture a large segment of the consumer market.

IDT charges customers a very low per-minute rate for long distance if the customer is willing to use its Internet access service. According to researcher Atlantic-ACM, the cost of acquiring a new Internet customer is $10–50. Therefore, IDT is able to use its price-differentiated IP telephony service to sell its commodity Internet access to new customers. BellSouth benefits from a similar bundling of enhanced services with a discounted Internet access package (about a 35% reduction). The enhanced services, such as call waiting, call forwarding, and caller ID, are unavailable from other competitive wireline providers in its service area, so these services provide differentiation. The large discount on Internet access is feasible because BellSouth benefits from the administrative efficiencies and the high profitability of the package of central office-based services undoubtedly help the overall return to BellSouth.

13.6 Supporting tools

Pricing innovations have moved more market share between service providers than most other strategies. Tools can assist the telecommunications planner to break away from traditional assumptions and structures and develop profitable, desirable, and proprietary pricing models.

Yield management

This form of revenue management is commonly practiced by airlines, an industry that often serves as a model of transforming markets through deregulation. Yield management tools enable service providers to set pricing algorithms that separate the price-sensitive buyers from those that are willing to pay higher prices. In the airline industry, yield management allocates seats per flight and then practices price discrimination to achieve management's objectives. The objectives can range from maximizing profit to increasing market share to balancing service utilization. The underlying parameters lend themselves well to many service industries, including telecommunications.

In the past, pricing has acted as the primary tool to manage network utilization and revenues. Nonetheless, price developers have only used devices that were intuitively or analytically available. Yield management techniques, especially coupled with data warehouses, can help telecommunications service providers develop innovative price structures to achieve corporate objectives in overlooked ways.

13.7 Role of information technology

Yield management, like other tools, requires hardware and statistical software to handle the iterative nature of the decision process. Eventually, yield management can move from its present location in corporate headquarters into a negotiation process with customers on the Internet. For example, a telecommunications service provider can offer to waive the installation fee if the customer is willing to wait several weeks before the service provisioning takes place. More sophisticated technologies could eventually set the negotiation parameters instantaneously. Only when the service provider develops a backlog, the discounted offer could surface.

Information technology has also facilitated the growing bandwidth commodities market, which could not exist without split-second transactions and instantaneous intercompany communications. Brokers can match buyers and sellers after they agree on a price for desired network minutes or take a larger role and manage the administrative components of the transaction—from billing to ensuring that quality-of-service measures are in the contract. Bandwidth commodity markets foster competition, because they give small service providers or resellers access to inexpensive sources of bandwidth that would not be available to them in conventional contracts.

One element that will assume great importance in pricing strategies is billing, and sophisticated technology is crucial to its success. First, measuring each call to include the elements of time of day, origination point, and minutes is complicated. Furthermore, the continued reduction in the cost of bandwidth makes the cost of administration a more visible component and a target for elimination. Second, object-oriented software and relational databases enable service providers (at least the ones willing to scrap their legacy systems) to offer highly personalized pricing and change their price structures quite suddenly, an advantage in intensely competitive markets. Service providers in countries without long-distance presubscription or preselection have created promotional pricing such as contests, events, or other incentives for casual calling.

References

[1] Knapp, J., "All-You-Can-Eat Long Distance," *Phone+*, Vol. 13, No. 1, pp. 76–77.

[2] Todd, K., "The Road to Local Competition," *Wireless Review*, Vol. 15, No. 23.

[3] Jenkins, H. W., Jr., "Wake Up Call for Bill Kennard," *Wall Street Journal*, Eastern Edition, Vol. CCXXXIV, No. 73, Sec. A, p. 31.

[4] Finnie, G., "Local Free-for-All," *tele.com*, Vol. 4, No. 5, p. 92.

[5] Montalbano, L., "It's on the House...Sort Of," *Phone+*, Vol. 13, No. 14, pp. 106, 108–109.

[6] Booz-Allen, "Customer Segmentation and Pricing," *Insights*, Vol. 3, Iss. 3.

[7] *Wireless Week*, "PCS 99: General News," Sept. 24, 1999.

[8] Britt, P. J., "No Easy Money," *Telephony*, Vol. 234, No. 23, p. 232.

14

Best-in-Class Performance

In theory there is no difference between theory and practice.
In practice there is.

—Yogi Berra

Many telecommunications providers hope and expect to differentiate their services based on superior organizational performance. In most technology markets, this is a feasible strategy. Intel maintains its leadership through its superior product development processes. Advertising campaigns such as Sprint's "pin-drop" and AT&T's "True Voice" demonstrate the desire of telecommunications service providers to differentiate their offerings through superior performance. Satellite communications provider Iridium's customers were often located in places without a great deal of supporting infrastructure. The service provider recognized that its customer care needed to be exceptional. Iridium created several virtual call centers, including language support in 13 languages, and access to support in 40 others [1]. Superior performance does not guarantee success.

In their book *The Discipline of Market Leaders* [2], Michael Treacy and Fred Wiersema identify three paths to leadership, which the authors call "value disciplines." The first, *operational excellence,* represents a service provider's ability to provide products at the best price with the least inconvenience. Practitioners of the second discipline, *product leadership,* strive to outperform their own previous superiority. *Customer intimacy,* the third value discipline, requires that service providers concentrate on the needs of specific customers. Applying the concept of value disciplines to the telecommunications services market challenges some fundamental conventional wisdom. Some telecommunications service providers assume that operational excellence translates into commanding high prices and high profits. While high profitability is indeed likely, operationally excellent companies generally accomplish their margins with lower prices than those of their competitors. Their superior business processes result in efficiencies and low costs. The best examples of operational excellence are uncontested market leaders, and in most markets, that means mid-market customers and low prices.

Superior operational performance does not have to mean only network reliability, response time, or transmission quality. AOL is an example of a service provider that won its customers through superior execution of its business model. When only a few major ISPs were operating, AOL's major competitors were Prodigy and CompuServe, both owned and well funded by major consumer-focused corporations. AOL, a startup, built its enormous user base on its ease of use, eliminating the need for customers to cope with a separate browser, e-mail client, and access provider. It also created relationships with branded content providers before the Internet became a primary destination for on-line customers. None of the elements is exceptional compared to its individual competitors, but their tight integration eliminates the need for users to have any technical knowledge. As AOL built its own community, it created switching costs for its customers and created an attractive mass of users for advertisers and new subscribers. AOL also pioneered low unlimited subscription prices while it maintained its first place lead among ISPs. This is a strategy for operational excellence.

Sprint has invested heavily in its networks and its external communications to meet its goal of product leadership. Both its long-distance "pin-drop" campaign and its wireless campaigns emphasize the quality of the network connection. Will this strategy work in the long term? That will depend upon several still unknown factors: How much of the network infrastructure will Sprint control on both ends of the call? Will packet-based telephony equal or exceed the quality of today's networks? Will another merger sacrifice either the brand or the supporting investment? Will a perceptible differentiation in network quality command enough of a price premium to cover the added investment?

Similarly, for customer intimacy, some telecommunications service providers strive to maintain customized, one-on-one relationships with a large portion of the universe of customers. On the contrary, the best examples of customer intimacy support a hand-picked customer base and make a significant investment in acquiring and keeping each one. Neither best-of-breed nor best-in-class signifies an unqualified best for the universe of customers. Companies that succeed based on preeminence are best within their own specifically defined market of targeted customers.

Treacy and Wiersema singled out the U.S. division of long-distance provider Cable & Wireless as a company focused on customer intimacy. The service provider specializes both vertically and horizontally. Knowing that it could not compete with the leading U.S. IXCs on price, it differentiated its customer support and created a knowledgeable sales force. In its horizontal orientation, it restricted its focus to business customers with monthly telecommunications bills in a specific range. Companies that did not support an in-house telecommunications staff were happy to call on Cable & Wireless for technical support. Its vertical orientation created a sales staff and in-house developers who created, sold, and supported services for targeted industry markets.

14.1 The challenges of preeminence

Competing through preeminence is appealing for telecommunications providers, partly because they have the resources to accomplish operational excellence and because of their full-service heritage. Nonetheless, there are many challenges to achieving superiority and earning returns from it, including the following:

- *For many telecommunications transactions, the service provider does not control (through outright ownership or outsourcing) major components of the customer's experience.* Most calls presently require the participation of several service providers; even a deregulated, bundled world of supercarriers offering every conceivable feature to customers could well require the assistance of another service provider to terminate a call. For instance, many customers are unaware that wireless calls travel for the most part along wireline facilities. IXC customers are quick to compliment or deride the network quality of their own service provider without considering that the local facilities at both ends of the call are often in the control of unaffiliated service partners. In the future, service providers that bundle local and long-distance services will control a majority of the customer experience but still not all of it in many transactions. Nevertheless, the service provider preparing to invest in the world's finest network needs to evaluate its benefits within the context of its reach.

- *Customer expectations for reliability and network quality are very high, even for the lowest-priced, least-differentiated providers.* Telecommunications service providers historically offered very high-quality service in most countries for many reasons. First, returns for regulated service providers often depended on their network investment, so this encouraged service providers to invest in their service quality. This is one reason that observers suspect that costs for incumbent monopoly providers are higher than they would be in a competitive market. This is not to imply that network investments were frivolous, simply that they were guaranteed a return. This is not the case in a competitive industry. Second, network investments that upgraded the quality of networks resulted in reduced customer service complaints. This reduced costs of customer service, and it reduced complaints to regulators, resulting in either enhanced revenues or at least a reduction in fines. The consequence of this high performance is that customers simply expect excellent reliability from the telecommunications network. As networks become more central to work, entertainment, and shopping, consumers will be even more demanding of service quality. This has already occurred in the business market. Therefore, the worst-but-surviving telecommunications provider will be a high-quality carrier. The best provider will need to exceed the standard of the worst performer and the average performer visibly to the customer.

- *Price remains a more important differentiator than superior service quality in many markets.* Customers do demand a quality baseline that is high, but price is still a powerful reason for customers to select and switch carriers once the baseline is achieved. Surveys often show that customers, especially consumers, value factors such as service quality, support, and brand. Still, they most often switch service providers due to favorable prices. Market researcher Claritas Inc. concluded from its survey of American consumers that for both local customers and long-distance customers, small discounts represent enough of a reason to switch providers [3]. For a discount of less than 10%, 22% of local customers would switch, and 43% of customers would switch for a discount between 10% and 20%. Nearly two-thirds would switch for a price reduction of more than 20%. The survey showed similar results for long-distance customers. One reason that prices remain a significant differentiator is that service quality is not yet a visible differentiator by itself. Thus, the dilemma for service providers is whether to invest in differentiating service quality in the belief that customers will change their behavior to their already-stated preference of better service.

- *Process excellence is readily reproducible by competitors, and potentially at a lower price than the innovator's.* Best-in-class performance can arise in any aspect of operations, not only the quality of the network. Service providers can achieve best-in-class operations in their execution of the distribution process, customer service, billing, provisioning, or a host of other business processes. Maintaining superiority in virtually any category will become a challenge for the leader, whether the performance leader is a new entrant or an incumbent. New entrants that bring innovative business processes to the marketplace will be hard-pressed to maintain their competitive edge against incumbents recreating their process designs, many of which cannot be patented. Incumbents with superior business processes will undoubtedly provide the benchmark for new entrants, so they will need to move quickly to improve at a faster rate than competitors, even when they begin with world-class performance.
- *Areas in which best-in-class performance is achievable must match the customer's criteria for buying.* Before the service provider decides the areas upon which to focus and invest, it must be certain that the area represents a buying criterion for customers that overcomes any price increase associated with the costs of excellence. This decision might be easier for service providers that have made a commitment to a specific market segment, with well-understood unique needs. For a specific segment, the value of customized services could exceed the small difference in price available in the general marketplace. For most customers, though, a superior business process is an incentive but not necessarily a compelling reason to stay with or switch to a

specific provider. Furthermore, sources of excellence that translate into price increases might not provide competitive advantage.

- *The pace to become best-in-class is hectic without a guarantee of success; the pace to maintain it is still demanding, but essential.* When a service provider that is not already in first place chooses to differentiate based on superior performance, the first task is to shrink the gap between present performance and desired performance. Service providers must accomplish this even while the best performers pull ahead at their own pace. Figure 14.1 demonstrates the process.

  This implies that the service provider that lags the best performer needs to improve at a significantly faster pace than the service provider that has already demonstrated its superior capabilities. Furthermore, the performance leader will not passively allow the challenger to win its first-place position, so the issues of competitive parity will surface.

14.2 The requirements of preeminence

While each organization maintains unique performance goals, best-in-class performance typically shares the following characteristics across organizations:

- *Executives actively promote efforts for improvement with words and actions.* Performance improvement initiatives by definition require

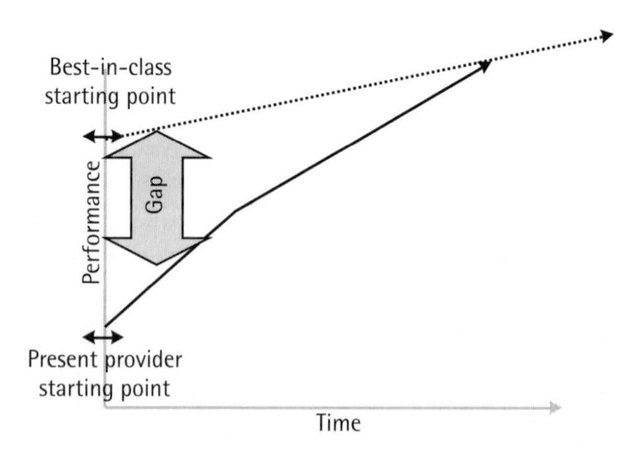

Figure 14.1 Shrinking the performance gap.

significant change to habitual business practices. When management acts as a champion of change, it offers employees both an affirmation of the organization's commitment to change and an implicit reduction of employee risk as lessons are learned during the improvement process. Management's commitment cannot end with the authorization to proceed or the acceptance of recommendations; senior management needs to embed superior performance in the core values of the organization.

- *Best-in-class organizations involve every member of the enterprise and, in many cases, the supplier chain in meeting performance goals.* Communication is vital in the drive to meet performance targets. Bell Atlantic's reengineering effort coincided with the launch of its intranet, and the communications network fostered the business process redesign effort. As telecommunications service providers depend more on outside suppliers for mission-critical activities, they will act as virtual employees and require the same infrastructure support. This requirement will become progressively more challenging for new entrants to handle, as their businesses grow and the number of internal relationships grows exponentially.

- *Best-in-class organizations maintain a sense of urgency to meet performance objectives.* In this area, new entrants often have a cultural advantage over incumbents. Monopolists control the overall pace of their businesses, not only the introduction of technology. Accordingly, former monopolists tend to be less inclined than entrepreneurial newcomers are to rush any process, even the process of continuous improvement. Because this is a cultural requirement more than an activity requirement, management must demonstrate leadership to employees, customers and suppliers. Often, significant business disruptions mark the moment that an organization decides to launch a performance improvement initiative. Indeed, the reengineering efforts of the RBOCs coincided with the imminence of competition in their local markets.

- *Strategic requirements drive performance metrics.* Performance tends to migrate toward meeting measurement criteria, sometimes at the expense of other tasks that are not measured. In addition, many temptations exist within organizations to measure the wrong dimension. One reason is that organizations sometimes use data collected in the normal course of business instinctively as the performance metric. Another reason is that some measurements are intuitively correct and are never tested for actual correlation with superior performance. If the performance objective is customer satisfaction with MTTR, one straightforward measurement is the lead time between the initial trouble ticketing and the completion of the repair. Service providers could use this measure indefinitely without recognizing that the goal is primarily about customer satisfaction, not internally defined processes. Did the customer call three different numbers over several days before reaching the employee that wrote the trouble ticket and launched the timer? Did the service provider notify the customer immediately when service was restored? Did the customer perceive the process to be professional and efficient?

- *Best-in-class organizations view the improvement process as a journey, not a destination.* The program is iterative, constantly reviewing business processes against changing market conditions. Routinely, best-in-class organizations hone the metrics, either to tighten their requirements to maintain their lead in a competitive market, or to revise or abandon metrics in favor of new measures that improve on the original ones. One method for ensuring that the process continues is to incorporate a continuous improvement program into the normal planning cycle. Another technique is to incorporate a best-in-class focus into the corporate culture, through a steady stream of value statements, corporate collateral materials, and other reinforcing communications.

- *Best-in-class organizations establish accountability for performance and provide the tools to reach performance targets.* Organizations improve the probability of success when there are consequences for poor performance and rewards for meeting objectives. Nonetheless, the service provider must ensure that the participants in each business process are equipped to handle its requirements and meet management's expectations. Resources to create an adequately supportive environment can include training, tools, and empowering employees to make decisions on their own that affect the operations for which they are accountable.

14.3 Case studies

Virtually every incumbent U.S. provider underwent a business process reengineering initiative (like many other U.S. companies) in the mid-1990s. For local service providers, the imminence of competition provided added incentive to reduce costs and improve service quality. Reengineering was especially useful to previously regulated providers, because its fundamental principles revolve around the customer. Monopolists tend to focus on regulators on the outside and their own measurements on the inside; customers are sometimes a secondary priority. Centering business processes on customers is time-consuming and complex but worthwhile.

Bell Atlantic, Pacific Telesis, NYNEX, and AT&T were among the service providers with aggressive reengineering programs with a goal of performance improvement. Among the targets for redesign was the billing system, the core of which for many of these providers originated centrally at AT&T before the 1984 divestiture. The flaws and inflexibility inherent in billing processes were illuminated squarely when MCI's Friends and Family's program helped move market share from AT&T to MCI. Billing systems are a growing market for third-party developers and are a viable candidate for outsourcing. The global billing market is estimated to grow to $15 billion in 2005 from its 1998 size of $3 billion [4]. Its 25% compound growth rate is more than double the expected rate for the telecommunications industry as a whole.

Service providers have recently focused on reengineering their provisioning processes for several reasons. First, the rise of the Internet has created an opportunity to reduce costs significantly and improve the cycle time. Across the industry, error rates within the provisioning process as high as 50% and more are common [5]. While these rates should be sufficient to alarm monopolists, they will undoubtedly terrify competitive service providers. Second, technology-based competitors have created provisioning applications that draw customers virtually on their own merit. Covad's customers can add or upgrade DSL lines from its Internet portal. Level 3 provides flow-through provisioning to its customers. CLEC Allegiance Telecom has chosen to develop its own standards-based, open operations support infrastructure and rely on it as a competitive differentiator. For the remainder of service providers, competitive response to actions like these is essential. Last, legacy systems need to be integrated to eliminate manual processing between steps and improve performance. In 1998 alone, service providers spent more than $2 billion on OSS integration [6]. Telecommunications service providers recognize that their customer service staff needs accurate access to order entry, billing, customer service, and provisioning history at least to maintain parity with the customer on the other end of the conversation. Moreover, for incumbent local service providers that offer wholesale network services, adequate and indifferent provisioning systems are the primary barrier to lifting the regulatory restrictions against offering long distance. The faster incumbents can create capable systems, the faster they can serve desired new markets. One added advantage for incumbents that plan to compete seriously in wholesale markets: Provisioning is undoubtedly a differentiating business process.

Bell Atlantic took advantage of its mergers with NYNEX and GTE by developing a best-practices merger philosophy. In the integration process, the company reviews both methods for conducting the business processes involved in OSSs, then selects the one that works the best to serve the customer [7]. Ameritech leveraged its 20% investment in Bell Canada to gain a window on its operations, as it did with its investment in 14 European carriers. The company sends a team of key people to compare the business processes of the target service provider with its own, then applies the best practices uncovered on a broad scale. Allegiance Telecom has taken an innovative approach to the development of a best-of-breed OSS solution. It uses its own operations as a testing ground for operations support. Allegiance believes that this approach improves the quality of the solution and speeds its time to market. Fixed-wireless provider WinStar created a "best practices training" workshop, which it offers at its branch offices. Each four-hour session focuses on a single sales topic in an effort to capture and disseminate sales knowledge throughout the organization.

Consultancy PricewaterhouseCoopers has identified practices associated with service providers that have reduced churn from as much as 35% to 20%. These practices include maintaining a well-integrated customer care program that gauges customer satisfaction with network services. Furthermore, the successful telecommunications service providers track and monitor customer behavior to target those subscribers that are likely to churn. Last, service providers with low churn sustain an organizational philosophy that emphasizes customer retention over customer acquisition.

Concerned about burgeoning local service competition, GTE recognized that its customer service lagged that of its competitors. The service provider undertook a benchmarking study in which it analyzed the practices of companies with superior customer service. The analysis revealed that customer churn was partly due to the customer's mistaken belief that competitors had better prices. GTE developed a tool to match customers with the most advantageous rate plan and reduced its churn in the study area by more than 20%.

Wireless provider AirTouch Communications developed a breakthrough customer care system that improved service and reduced response time. The system included features such as exact bill image representation, bill recalculation, price planning, service changes, and other features before many of its competitors had upgraded their own systems. The initiative involved many team members from senior management to supervisors and users. The teams defined business processes that integrated cross-industry best practices and incorporated them into the newly designed process. The resulting systems reduced training time for service representatives by more than a week. Eighty percent of customer calls could be handled on the first system screen. On-line rerating reduced the time to recalculate a customer's bill by 75%.

Sprint's customer service benchmarking effort revealed that its best-in-class partners did not necessarily maintain a voluminous set of practices typical of a regulated telecommunications service provider. The benchmarking process challenged other conventional wisdom; Sprint learned to look inside its own organization first for best practices before looking to external partners.

14.4 Partnerships

In 1996, AT&T and Pacific Bell sponsored a study of product development to create usable metrics and a better understanding of best-in-class performance [8]. The objective of the effort was to measure the difference between high and low performers and to identify the practices that cause a service provider to become one or the other. More than 30 wireline and wireless service providers in the United States, Europe, and Asia participated in the initiative. The study concluded that there are significant measurable differences between the highest performers and the average performer. The companies with best-in-class time-to-market performance delivered services from concept to launch in half the time of the average performer. Average wireless development took 21% less time than average wireline development and wireless time-to-market was 36% faster than wireline. This discrepancy could portend some of the competitive dynamics as wireless substitutes more for wireline service. The study also showed that best-in-class companies set larger improvement goals for themselves, which, if achieved, would further widen the gap between the best performers and the average performers.

More than 250 telecommunications service providers are active in the TeleManagement Forum, a nonprofit global organization that focuses on improving the management and operations of communications services. The organization facilitates the development of standards, performance benchmarks, and interfaces between service providers to ensure a seamless telecommunications solution for customers. It also conducts proof-of-concept demonstration to illustrate the deployment of technology.

The Alliance for Telecommunications Industry Solutions (ATIS) assists its members in developing network interoperability standards in areas such as interconnection, numbering, toll fraud, network reliability, T1, and other technologies. Twenty-five service providers formed the Applications Service Provider Industry Consortium. Its goals include education, common definitions, research, standards, and the identification of best practices. Dozens of other industry groups maintain benchmark data or best practices definitions for the use of members. Even supplier Sun Microsystems supports a SunTone certification and branding program. Its intention is to develop a set of best practices for the service provider industry in hardware infrastructure, security, and operational processes. Certifications such as these will benefit new entrants, which will need to convince customers that mission-critical applications will sustain a performance level that rivals the incumbent's brands.

In 1999, Australia's Productivity Commission published an international benchmarking study comparing Australian telecommunications services to other world service providers [9]. The study compared competition policy, prices (Finland and Sweden showed the lowest prices for most services), and a variety of quality of service indicators. The study discovered that countries that moved fastest to competition generally supported the lowest prices, and that the best performers had prices 20–40% lower than Australia's. The study also determined that internal as well as external factors explain its differences in performance. Therefore, the study concluded that a combination of government policy and management attention could improve the telecommunications service environment for customers.

14.5 Supporting tools

Many opportunities are available for service providers to create distinction through technology. For provisioning, third-party workflow software is available, improving, and quite competitive, due in part to the rising market interest in improving performance.

Benchmarking

Benchmarking has become popular in the last two decades, because it enables companies to evaluate their performance in the context of their peers. The company most frequently associated with the new benchmarking model is Xerox, the inventor of Ethernet and a significant if underrated participant in the telecommunications industry. Benchmarking applies to many aspects of corporate performance, but it lends itself especially well to operations, because measurement is less subject to bias and variation than other business processes. Benchmarking itself is a tool, but it can become a path to differentiation when it signifies meeting customer expectations for service quality.

Benchmarking is not a new management tool, but in the last decade its practitioners have honed its techniques to be more robust than in the past. AT&T and its local subsidiaries maintained the huge volumes of Bell System Practices, which aggregated the most effective business activities throughout the company and recommended that all employees follow the most successful examples.

There are three distinct types of benchmarking: process benchmarking, performance benchmarking, and strategic benchmarking. All three are useful points of departure for telecommunications service providers. Process benchmarking looks at specific business processes, such as billing, provisioning, or repair. This type of benchmarking has produced dramatic results throughout the business community by eliminating large amounts of work through automation or innovative process redesign. Cost reductions of 50–80% are possible when benchmarking augments a conventional reengineering effort. Cross-industry analysis can help the process benchmarking effort. In the course of reengineering eight core operational processes, GTE examined the practices of more than 80 companies.

Performance benchmarking compares competitor services to one's own in terms of price, quality, features, and other characteristics. For most products and some services, reverse engineering is a useful technique to analyze the costs of a service and the reasons for its success. For services that do not lend themselves to reverse engineering, service providers can use mystery shopping programs and direct analysis of operating performance measures.

Strategic benchmarking is the least quantitative of the three types, but it can nonetheless have a significant impact on operational and corporate performance. Most often, strategic benchmarking is cross-industry to uncover those strategic practices of superior performers that can create similar performance in one's own organization. Strategic benchmarking holds a long-term view of corporate transformation and therefore does not ordinarily produce immediate results. Still, strategic benchmarking provides an important perspective for a forward-thinking organization.

Enterprise resource planning

ERP packages are a usual information systems initiative in most corporations. Besides improving operational performance over time as most information systems projects do, implementation of ERP forces organizations to review all business processes even before they design the new systems. ERP software, also called enterprise resource management (ERM), manages an organization's information systems along the requirements of its primary business processes. One benefit of ERP implementation is that these systems integrate the service provider's data and make it accessible in a usable format to employees all over the organization. ERP systems often achieve a large percentage of their own savings by minimizing the maintenance costs, which are often considerable in legacy in-house systems. Many service providers installed ERP systems in the last years of the 1990s to eradicate the threat of Y2K noncompliance from their legacy applications. Thus, the requirement to solve an imminent business problem had the favorable consequence of creating a strategic benefit at the same time.

Deutsche Telekom uses a combination of purchased software and systems developed in-house to meet its ERP needs. Integrating diverse systems is challenging, and the service provider depends highly on the use of standards to ensure system compatibility. Telekom Austria implemented enterprise resource systems for several of its financial processes. Its investment in these systems achieved its payback a full year earlier than planned. Swisscom migrated from an in-house system toward a purchased ERP solution in 1998 and anticipates that payback will occur in two to five years. WinStar needed a flexible system and a rapid time frame for implementation when it designed its ERP system. Its system also integrates the information system requirements of WinStar's acquisitions, which is key to the service provider's growth strategy.

14.6 Role of information technology

Information technology has been the foundation of virtually every performance improvement effort in the last two decades. The large data analysis requirements for benchmarking studies and other performance improvement initiatives require state-of-the-art analysis and frequently statistical tools. ERP implementation is among the most significant information technology decisions that a service provider can make, and much of the cost-benefit justification will rely on potential savings or expected service improvements.

Within the network infrastructure itself, information technology, coupled with the rise in packet-switched Internet-based networks, will undoubtedly drive best-in-class performance for every business process and most innovative services for decades to come. This will most likely increase the priority of supplier relationships or vertical integration for service providers. Vertical integration will be a necessity for service providers with unique requirements or the need to own technologies without the participation of vendors. This is a significant strategic decision, and it is unlikely to predominate in a highly competitive environment. The more likely arrangement is that service providers will develop significant alliances with leading equipment manufacturers or with superior niche manufacturers or software developers. Telecommunications service providers will help fund the development of systems or hardware in exchange for an exclusive license to use the resulting technology. Performance can be both best-in-class and unassailable by competitors.

References

[1] Grambs, P., and P. Zerbib, "Iridium Customer Care," *Telephony*, Vol. 235, No. 19.

[2] Treacy, M., and F. D. Wiersema, *The Discipline of Market Leaders: Choose Your Customers, Narrow Your Focus, Dominate Your Market*, Reading, MA: Addison Wesley Longman, 1995.

[3] Turner, T., "Does Convergence Cause Churn?" *tele.com*, Vol. 3, No. 9.

[4] "The Importance of Billing Earnestly," *America's Network Telecom Investor Supplement*, Vol. 103, No. 18, p. 6.

[5] Kohn, K., "The Incredible Shrinking Interval," *Telephony*, Vol. 237, No. 18, p. 58.

[6] Davis, J., "Beating Those Post-Merger Blues," *America's Network Telecom Investor Supplement*, Vol. 103, No. 12, pp. 12–16.

[7] McElligott, T., "Multiple Personalities," *Telephony*, Vol. 238, No. 5, p. 32.

[8] Ogawa, D., and L. Ketner, "Benchmarking Product Development," *Telephony*, Vol. 232, No. 4, p. 35.

[9] Productivity Commission 1999, *International Benchmarking of Australian Telecommunications Services*, Research Report, AusInfo, Melbourne, March.

15

Customer Service

My own wife told me, "You're in charge of billing, and the phone bill is the dumbest bill I get."

—RBOC billing executive

Good customer service, or its new nomenclature, customer care, requires that a service provider's customer have a positive experience each time there is contact between the service provider and the customer. Like many other measurement parameters, good service is necessary but not sufficient for differentiation. Most service providers will find ways to keep their customer service at that level. Service providers that intend to differentiate their services on the basis of customer care—and that includes the stated intentions of most of the industry leaders—will need to go much further, providing unmistakably superior customer service. Furthermore, differentiating based on customer service means keeping more customers who stay simply for the customer care experience. Therefore, customers will need to be willing to forego discounts and other enticements from competitors because of a loyalty to the customer service of their existing providers. Telecommunications service providers that commit to differentiation in this area will need to be up to the challenge.

15.1 Service as differentiator

Customer service offers more opportunities for a service provider to differentiate from its competitors than most other paths to differentiation. Figure 15.1 depicts some of the opportunities that occur throughout the entire customer life cycle.

In the customer acquisition phase, the service provider can differentiate through its sales team (or sales venue), order handling, provisioning, or fulfillment. In sales, a telecommunications service provider specializing in a vertical market can offer expertise to its customers. The sales team for a consumer products enterprise can hold industry knowledge, through sales training, through immersion in the industry press, or through industry experience. The team can assist the customer in analyzing its industry position in terms of its communications infrastructure. The sales representative for a small business can provide presales analysis of capacity planning and can design a sound network solution for a customer that is unlikely to have much technical knowledge. The solution, of course, needs to be proprietary to the service provider the salesperson represents, so customers do not have the option of taking the network design to a less supportive, less expensive provider. Preeminence in the sales process, by itself, can differentiate a service provider sufficiently against its competitors. Order handling, provisioning, and fulfillment tend to act less as inducements for customers to switch to a certain service provider and more as customer retention mechanisms once the service provider has demonstrated

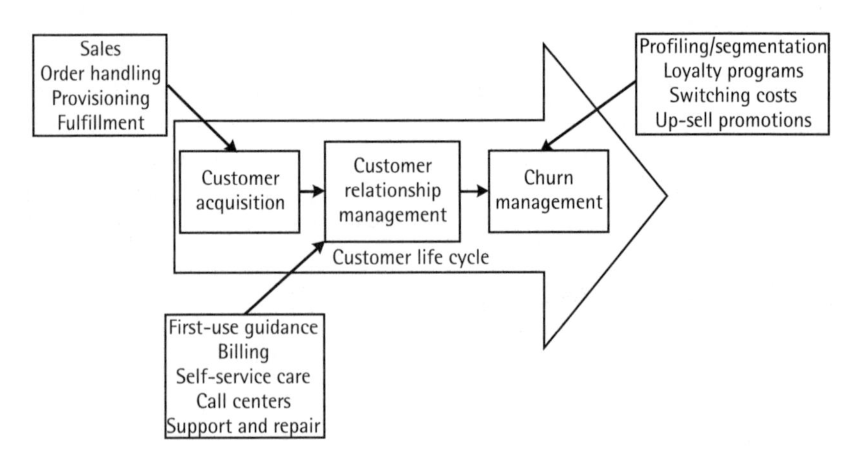

Figure 15.1 Customer service differentiation options.

capability in these areas. Indeed, terrible service experiences in these areas can be an incentive for a customer to switch to a service provider with a better reputation.

During the active management phase of the customer life cycle, customer service represents the only ongoing contact between a service provider and its customers. Opportunities to differentiate include first-use guidance, billing, self-service customer care, call centers, support, and repair. First-use guidance is presently unnecessary for generic local and long-distance services, but it is already important for complex enterprise networks, ISPs, and some niche services, such as prepaid services. This is the first customer contact, and it will create strong first impressions of overall service quality. Service providers that invest in this customer support will find that they can improve customer satisfaction, increase service usage, and reduce calls for customer support later. Billing is a most important component of customer service because of its frequency. Telecommunications providers have not traditionally taken advantage of the customer service opportunity in the bill. Similarly, telecommunications service providers tend to lag other high-technology organizations in moving their businesses to the Internet, with some noteworthy exceptions. Telecommunications service providers have used call centers effectively for decades, and the Internet and other technologies are creating new opportunities to enhance the customer's service experience. Billing, self-service on the Internet, and call centers are so strategically important that they merit separate discussions in Sections 15.3, 15.4, and 15.5, respectively. Like order handling and provisioning, support and repair have more potential to damage the customer relationship than to create a positive one. Still, telecommunications service providers that intend to differentiate based on superior customer service need to ensure that they protect their investment in potential differentiators such as billing with adequate attention to customer service requirements such as support and repair.

Customer service can act as a positive differentiator before a customer decides to churn or even after the customer considers leaving. Techniques to retain the customer through customer service include profiling and segmentation, loyalty programs, switching costs, and up-sell promotions. Segmentation can offer opportunities for customer retention. If a service provider targets a specific group or community, the service provider can customize the offer that less focused competitors cannot match. A small service provider can then compete successfully against a large provider. Suppose an emerging service provider decided to target an immigrant community. The service provider could build critical mass in the community by creating a highly differentiated service. The service could include an option for all communications and call center support in the language of the community. In addition, large calling volumes to the home country would command low prices from suppliers and lead to favorable international calling rates. The telecommunications service provider could maintain a Web portal with demographically valuable information and links, including news, community activities, and travel promotions.

Loyalty programs are an important part of customer service. The best customer service programs build rewards over time. This creates a large switching cost for the customer, and it guards against a competitor's ability to match the award for a new customer. For example, a service provider could create a loyalty program that offers some amount of free minutes of service after six months, double that amount after a year, and double again after two years. Switching costs create a reason for customers to stay and a cost for competitors to disentangle the customer during their own acquisition process. Many service providers do not exploit the selling opportunities available when customers call to state their intention to churn. Service providers committed to customer retention should know what competitive packages are available, including promotional inducements, and offer comparable services to churning customers to entice them to save themselves the inconvenience of churning.

One disadvantage to current long-distance presubscription (and to casual long-distance calling as well) procedures is that the customer can churn without notifying the incumbent service provider. The service provider's first notification occurs after the customer is already lost. The number portability guidelines for local service (and eventually for wireless) indicate that local service providers will have the same disadvantage. Deregulation makes churning as easy as possible. Therefore, service providers need to find ways to contact the customer before the opportunity to churn arises. If the service provider can reach the customer before the competitor does, there is an opportunity to recover the account. This is the reason that customer relationship management and constant customer contact is so important.

It is critical for telecommunications service providers to develop stores for relevant customer data and tools to ensure that the customer's contact has access to complete, accurate, and well-indexed information during each customer communication. Integrating applications for all business processes is important, but it will soon represent the lowest standard for customer service. Telecommunications service providers that intend to compete based on superior customer service will need to introduce innovative, if not proprietary, information retrieval capabilities.

There is another reason for incumbent RBOCs to focus on customer service—their role as wholesale provider. This reason is twofold. First, meeting the requirements of the 14-point checklist is what stands between the RBOCs and the ability to bundle local and long-distance services to their customers, an opportunity to reach a very large market. According to the Yankee Group, Bell Atlantic's in-region long-distance opportunity is $20 billion. (Bell Atlantic is only permitted to offer long distance to its customers in some areas, so it still has challenges to meet.) SBC can win part of its own $23 billion per year in-region market, including its PacTel and Ameritech territories, and BellSouth and US West are eager to gain their shares of their respective $8 billion and $6 billion markets in their regions. Customer service generally evokes the image of retail customers, but for incumbent LECs, wholesale customer care is a significant issue. First, CLECs can choose to let the LEC wholesaler provide customer care to its retail customers. Second, the same OSS infrastructure and other business processes that support retail customer care support their wholesale customers. Indeed, this is one reason that OSS is such a fundamental and problematic part of the 14-point checklist. US West's OSS test plan encompasses almost its entire territory, primarily to hasten the approval process for its long-distance applications.

Some service providers were unprepared for the second-line growth caused by the emergence of the Internet. US West thought that it would improve its returns without affecting customer service when it consolidated more than 500 service centers into 26 new facilities as part of a reengineering effort. At the same time, demand for second lines was building. The combination of new demand and a smaller work force led to customer service delays and failures [1]. GTE and Pacific Bell experienced similar service problems. Another service provider's management became aware that the marketing department was actively promoting second lines without notifying the network managers in advance.

The complexity of the decisions required about a service provider's commitment to—and investment in—customer service underscores the fact that no carrier will be competitive in every market. Customer service excellence is expensive. As in other areas, telecommunications service providers will need to decide to compete wholeheartedly or compete adequately.

15.2 Improving traditional forms of service

J. D. Power and Associates ranked Sprint (among high-volume residential long-distance users) and SNET (in the low-volume segment) highest in 1999. Sprint's ranking was the fifth year in a row, and SNET's was the third, so each of these service providers has found some core competencies that resonate with customers. According to J. D. Power, one of the most significant drivers for residential long distance is customer service. In contrast, the J. D. Power and Associates 1999 wireless customer satisfaction study revealed that customer service is not a primary influential factor affecting satisfaction with wireless services. For wireless customers, customer service trails call quality, pricing options, and corporate capability and leads only the cost of roaming, the handset, and billing.

TCI (now owned by AT&T) was among the largest cable providers in the United States, but it ranked last in a 1997 J. D. Power and Associates customer satisfaction survey. The cable provider immediately initiated programs to improve its ranking, which it accomplished before the next year's evaluation and its subsequent purchase by AT&T. One concern shared by cable operators was that the mix of cable-based services was becoming more diverse, to include telephony, broadband, and more video choices. Service needed improvement in anticipation of the customer confusion and the new skills required of service representatives. MediaOne (now owned by AT&T) created its Broadband University to teach service representatives about new technologies and launched a work-force management system. Jones Intercable learned that its customers were dissatisfied with the way service representatives sounded on the telephone, so it hired a voice coach for them. TCI invested in training and its billing system, but the service provider also reviewed its compensation structure to boost morale.

The Strategis Group conducts studies routinely to reveal the differentiators in a competitive telecommunications environment. In a 1998 version of its business branding and bundling telecommunications services study, the market researcher discovered that incumbent local and long-distance service providers have a competitive advantage over new entrants in the area of customer service and quality assurance. Among business users, 61% are highly satisfied with the quality of service from local companies, and 70% are highly satisfied with long-distance providers. Furthermore, the study provides affirmation that service can be a differentiator: 47% of business users cited better service through an integrated network as a very important reason for switching to a sole-source telecommunications provider. While the overall quality of the network is also a significant element in the customer perception of quality assurance, customer service is responsible for quickly responding to customer outages, integrating the customer's administrative experience, and gaining the customer's trust.

Bundling of services becomes an important customer service issue, because the customer is highly dependent on a single provider to manage its customer care. Most bundling studies demonstrate that consumers are eager to bundle their telecommunications services with one provider and often see their local or long-distance provider as the obvious choice. Nevertheless, the Yankee Group's technologically advanced survey of consumers found that fewer than half would rank the professionalism, courtesy, and knowledge of their local service provider as above average or excellent [2]. Only slightly more than half ranked long-distance providers as above average or excellent on the same parameters. This is especially important in light of J. D. Power and Associates findings that customer service was the most important factor in customer satisfaction with local service providers and the second most important factor in long distance.

One conclusion of the various J. D. Power rankings can support the implication that customer service will decline in importance as markets become highly competitive. Customer service ranks first in local markets, where most consumers do not have a choice of providers. Where choice is readily available, in long-distance markets, price is more important than customer service. In wireless markets, choice is available but network quality is not as robust as their wireline counterparts are. Customer service ranks behind service quality and price. One implication is that competitive telecommunications markets will drive customer service to a lower priority, especially if all service providers meet basic network needs and if services remain the near-commodities that they are. Another implication, though, is that once markets have stabilized and price competition subsides, customer service can emerge to act as a differentiator. Many service providers appear to be willing to wait.

15.3 Billing

Pricing of telecommunications services is becoming both simpler and more complex. Flat-rate offers greatly reduce the calculation and sheer bulk of a customer bill. E.spire moved to a flat-rate model largely to avoid the complications of a usage-based billing system. On the other hand, telecommunications service providers need to capture customer usage data and manipulate it in their data warehouses. Furthermore, bundling services and providing a single bill (one of the key reasons customers support bundling) means that billing systems need to accommodate a variety of formerly separate services. These could range from Internet access, to local and long distance, to equipment purchases, to usage-based data or video services, in any conceivable combination. Moreover, intense competition will create the need to change prices quickly, create promotions on short notice, and manage an array of customized service packages.

Billing data serves important marketing functions beyond simply calculating revenues. This is one reason that service providers would capture call data even if they offered all services at a flat rate. Usage data and account information for an individual customer enables the proactive service provider to recommend new packages or discounts that would reduce the customer's expenditure or head off churn before the end of a customer's service contract. While some service providers are choosing to develop their billing systems in-house, third-party software developers have risen to the challenge of creating usable, state-of-the-art billing systems that can generate and manipulate billing data. The billing market by itself will be an important service provider wholesale niche. Communications Industry Researchers (CIR) estimates that by 2006, telecommunications service providers will buy more than $10 billion in billing systems and services, compared to about $2.5 billion in 1997 and $3.2 billion in 1999. Another estimate of this market is that the global market for billing services will expand to $15 billion in 2005, from $3 billion in 1998—a 25% compound annual growth rate [3].

Another benefit of supporting excellent billing capabilities is that monthly bills are a branding opportunity. While it is true that many customers do not enjoy receiving bills, in general, they are receptive to other messages inside the envelope. Telecommunications service providers often use bill inserts to sell additional services. Still, the incumbent providers are not skilled so far at branding their services to consumers through their monthly bills. Fifteen years after divestiture, market researcher Insight Research determined that 5% of telecommunications consumers did not know who provided their long-distance service, and another 5% incorrectly identified their local service provider as their IXC. One reason for this confusion is that the long-distance bill often appears on the local service bill, creating a branding challenge for the long-distance provider. Of course, a very strong brand can overcome the persistent if bland local service bill. Ten years after giving up the local service market in its divestiture from the regional Bell companies, 30% of consumers named AT&T as their local provider [4]. This number dropped to 5% in a later survey.

The billing and payment process is an opportunity to connect with customers or to disappoint them. According to Insight Research, half of long-distance billing problems take more than 24 hours to resolve or are never resolved. Four categories account for 80% of the total problem statistics. While this demonstrates that telecommunications service providers have not excelled at expeditious customer service, it also implies that there is a differentiation opportunity for the service provider that can offer superior billing support to customers that care about billing.

15.4 Web-based service

Telecommunications service providers have enhanced their customer care and reduced their costs by moving certain functions to the Internet for customers that choose to transact business on-line. Many service providers offer on-line bill presentation and occasionally customers can pay their bills on-line as well. In addition to the customer's convenience, on-line commerce reduces administrative costs for the service provider. These costs can include the cost of printing and mailing the bill; the labor costs involved in opening bills, such as matching payments to amounts due; manually posting transactions to the accounts receivable system; and often the cost of float, the time between bill presentation and payment clearance. According to NIIT, the cost of printing and mailing a bill is between $2.75 and $4 per residential customer [5]. At its own estimate of $2.50 per month, Insight Research estimates that on-line billing for consumer telecommunications services will save service providers $36 million in 2000 and nearly $400 million by 2004.

The most basic Web-based customer care can include frequently asked questions or information about the company, such as a list of press releases. Telecommunications service providers can enhance this static content with an option for customers to send an e-mail message with a specific question. The e-mail can go the other direction as well. If the customer inquires about the availability of a new service for which deployment plans are not ready, the service provider can promise to send an e-mail with the schedule when it exists. The customer can ask a service representative to call or speak to a service representative on-line through text-based chat or live audio. Customers can transact commerce through bill payment, order entry, trouble reporting, and tracking.

One advantage of migrating customers to the Internet for self-service is the cost avoidance for the call center. Insight Research estimates that labor costs account for about 70% of a service provider's expenditures on billing and customer care. When customers answer their own questions on-line, the service provider avoids charges for toll-free calls and the service representative's time and resources, which can range from $3 to $11 per transaction, depending on the length and complexity of the call [6]. Service providers can thus reserve their more expensive call center resources for more complicated or atypical customer concerns. Besides the cost of developing the Web site, a less apparent cost is that the average remaining transaction directed to the call center will require more skills on the part of the service representative. Service representatives will need more training and potentially higher wages when more users are getting their routine requests resolved on-line.

Besides its ability to reduce the overall cost of providing service, Internet-based customer care has other advantages. The typical Web user is more attractive than an average customer to most service providers. Web users have higher telecommunications bills and use more high-profit services. Differentiating a Web-based customer service program will attract and retain these profitable customers. Internet sites have sales opportunities that are unavailable to most direct marketers and advertisers and compare well to the power of direct sales, the most expensive sales channel. On a Web site, the service provider can describe in meticulous detail services that are simply too obscure to sell through telemarketers or train retail sales representatives. An on-line catalog can assist customers in sorting through an abundance of services, at a fraction of the cost of a transaction with a customer service representative. Insight Research estimates that the e-commerce catalog market will generate $13 billion in revenue in 2000, or about two-thirds of the on-line commerce market. In addition, instead of the push of a sales call, Internet users—who know their own needs best—seek out the services that interest them. Internet sites offer unequaled branding opportunities for what might otherwise be a commodity service—if the service provider has designed the site effectively. A good site will hold the interest of a customer for longer than a television commercial and offers the opportunity to buy immediately. Internet investments also tend to replace the variable costs of most sales with the fixed cost of a hosted site. As the site attracts more customers, the average cost per transaction falls significantly.

Web-based support can create opportunities for customer dissatisfaction as well. One challenge that on-line support still faces is where to direct e-mail requests. On-line customers are often less patient than the general customer base because they are accustomed to getting answers quickly. When the e-mail requests are inefficiently handled, the customer's request can languish for days or longer. In 1998, Forrester Research estimated that 60% of companies manually rerouted e-mail customer contacts.

Telecommunications service providers were historically unwilling to allow customers to manage their own networks and were even reluctant to let customers view network management data. The availability and capabilities of Internet portals—and probably the ready availability of more forthcoming competitors—have changed their cultural caution to acceptance and even endorsement of Web-based network information. Systems that did not use the Internet often required customers to purchase dedicated terminal equipment and learn to use proprietary software. Security on the Internet is better than that of in-house system development, so the customers and the service providers are confident that the data is safe. Large customers have access to bill presentation, but they can also provision new services, track network faults and service repairs, monitor overall network performance, and perform virtually any function available to network managers overseeing their own facilities. This is especially important to resellers, which need timely information to meet the needs of their own customers.

AT&T's portal, Interactive Advantage, enables business customers to conduct network performance reporting, network control, ordering and status reporting, account information, trouble reporting, and reference information. The system handles more than one-third of AT&T's customer care transactions using dozens of tools, most of which incur no charges. AT&T's customers can order ports and check the status of frame relay, private line, and voice services. MCI WorldCom's portal Interact includes service management tools, invoicing, order entry, customer care, and personal productivity tools. MCI enables its customers to add bandwidth dynamically to existing private lines and ATM links from a Web browser. Web-based Qwest Control offers trouble management, detailed statistics, configuration, and billing capabilities to customers. Sprint's service is available to its ION network customers.

15.5 Call centers

Call center technology has improved at the same pace as other communications technologies. Telecommunications service providers frequently offer call center services to their customers, but their own call centers are not particularly outstanding. One possible reason for this is that most telecommunications services are featureless. A customer picks up the telephone, receives dial tone, and makes a call. For normal communications, customers rarely need help. For customer service problems, most telecommunications service providers have traditionally viewed the call center as an expense, not a service opportunity. This will change as communications services become more complex. Microsoft uses its call centers as a point of differentiation in a highly complex user support environment. When customers call Microsoft's software support line, it is often the case that a technical representative is not available. Customers are required to hold, often at long-distance rates, as the support lines are not toll-free. Microsoft recognized that the customer calling for Windows support is most likely already unhappy and that the hold time waiting for the support technician can make them even more dissatisfied. The software developer enhanced the customer's wait with a disk jockey and music on hold, and between performances, a voice-over announces the maximum wait time for each product. The customer can then decide whether to wait on hold or call at another time. By the time the support representative picks up the call, the customer is more relaxed and is in a problem-solving mind-set.

Wireless provider LA Cellular (now owned by AT&T Wireless and BellSouth) decided that customer care was an effective way to differentiate in the crowded Los Angeles cellular market. Its call center screen pops (information shown to service representatives based on caller ID) immediately verify if the customer is eligible for support, reducing on-phone time by about 20%.

Telecommunications service providers practicing customer segmentation can use call centers to provide individualized service to customers. Caller ID enables the call center to route the call to the most capable service representative for the anticipated customer service request. A service provider specializing in a geographical or demographic market could offer native-language customer support. The call center software can route the caller whose records show a recent contact to the same service representative that handled the original problem. If a customer just installed a new service, such as DSL, the call center can route the call to the technical support staff member specializing in that service even before the customer states the problem to a service representative.

Web-enabled call centers are becoming more common, and video transactions will undoubtedly replace audio and chat once more customers are connected at broadband rates. Web-enabled call centers combine the call center transaction with the on-line session. This can enhance the customer's experience through instant messaging, e-mail, synchronized sessions, and click-to-call, in which a customer clicks an icon on the Web page and instantly connects to a service representative.

15.6 Customer relationship management

Customer relationship management (CRM) is the discipline of finding, attracting, and retaining those customers that support growth and profits. CRM has expanded in importance in many industries, especially service industries, largely due to increased competition on the demand side and the availability of useful technical and decision support tools through information technology on the supply side. These tools have made it possible to quantify the impact of customer relationship management on profitability and success. Like other strategic initiatives, knowing that CRM can improve the business is not sufficient; management must ensure that the investment in CRM returns more than its expenditure and more than alternative uses for the investment. Andersen Consulting determined that CRM performance accounts for 50% of the variance in return on sales for service providers [7]. Furthermore, Andersen identified that only 12 of the 54 capabilities it studied drive most of the ROS variance. Of the 19 key customer service capabilities, three alone—an effective billing system, attracting the best talent, and measuring customer service effectiveness—accounted for more than 43% of the impact of moving from average to top-tier performance in customer service. Andersen further calculated that a $2 billion service provider that improves its overall CRM performance (including marketing, sales, and customer service) from average to top tier could increase its return on sales by more than $320 million.

Data mining is useful as an element of a customer relationship management program. These heuristic analysis techniques can often detect trends before they create churn. If a customer's usage is falling while prices are decreasing, it could mean that the customer is in economic difficulty, or it could mean that the customer is using another service provider. Either way, a call from a service representative can help the customer create a new payment plan that benefits both the customer and the service provider, or it gives the service provider the opportunity to recapture the customer's usage before churn occurs. It might be that poor network quality has driven the customer to an alternative provider for important calls. Without data support, these facts would not emerge until the customer churns, and maybe not even then. Active solicitation provides excellent customer feedback and can reduce churn or increase service revenues.

15.7 Supporting tools

The tools supporting customer service augment the telecommunications provider's operational measures, which sometimes depict an incomplete view of the customer's perspective. Using quantitative methods, the strategic planner can examine the actual rather than the expected viewpoint of the customer. Furthermore, predictive tools can assist the telecommunications provider in reducing churn.

Customer retention analysis

A customer retention analysis typically aggregates information about the length of the customer's life cycle, life cycle profitability, churn rates, and repeat customers. Once this information is gathered, analysis can demonstrate profiles of those customers who offer the most profitability over their life cycle, characteristics of customer loss, and customers who have untapped potential for additional services. If the telecommunications service provider supports a data warehouse, this analysis is a typical application. If not, any service provider can perform a customer retention analysis using sales data and either a spreadsheet or preferably a relational database program.

Customer satisfaction measurement

While many metrics work well to evaluate a telecommunications service provider's performance as compared to its objectives, customer satisfaction measurement adds an important dimension. This tool enables the service provider to link its internal measures with an external consequence, the satisfaction of the customer. In some ways, this analysis acts as a measure of a measure. Customer satisfaction measurement also acts as an early warning of churn and a robust if challenging feedback mechanism. Areas for improvement identified in this analysis need to command a high priority for resources to correct them. Customer satisfaction results also provide valuable longitudinal data. If customers are becoming less satisfied over time, it signifies either that quality has decreased or that customers are becoming aware of attractive competitive alternatives. Customer satisfaction research also provides an opportunity to ask customers about their own service priorities, so it is among the most direct analysis tools that exist. Moreover, this tool can be a part of a benchmarking program, to compare one's own performance directly against that of competitors.

15.8 Role of information technology

Technology has a significant impact on customer care, whether the service provider commits to a large investment or not. The more automated the service provider's underlying business processes are, the more accurate the data, creating fewer customer service transactions initiated by the customer. With more technology, customer-facing systems will have more capability. Furthermore, more automation results in more standardization, reducing staff training costs and creating an architecture that customers understand readily. Last, information technology most often results in cost effectiveness. Therefore, whether customer service is face-to-face to the customer or behind the scenes, information technology is a sound investment for most service providers.

As technologies have improved and software applications have diversified, customer relationship management software has developed in most industries. CRM systems integrate customer information from diverse corporate systems and provide customer service and sales representatives with indispensable data before and during the customer contact. A CRM system can support the customer life cycle from lead management through sales through customer retention. Similarly, call center technology is a cornerstone of modern customer management.

OSSs are rising to meet the challenges posed by customer care requirements. In some ways, incumbents have an advantage over newcomers, because they already own systems that work with their networks. On the other hand, new entrants have their choice of new information technologies without the burden of legacy system maintenance. Furthermore, the third-party OSS market is growing at 30% per year [8], compared to overall market growth of 10%.

OSSs need to integrate a variety of business processes, including order entry, provisioning, billing, and repair. For incumbents, these systems were historically completely separate. Mergers combining decades-old systems create more complexity. New entrants often have the advantage of the newest platforms. Still, resellers rely on the operations support of their suppliers (often more than one) and facilities-based competitors often outsource large portions of their service portfolios. Thus, even the new entrants sustain integration challenges. OSSs can prevent customer service failures as well as provide support during operations. A facilities-based telecommunications service provider can isolate network failures, restore service, and notify the customer, sometimes before the customer is aware of the problem. This proactive use of technology has the potential to become a baseline requirement for customers.

References

[1] Lawyer, G., "Leaner, Meaner—and Busier," *tele.com*, Vol. 3, No. 4.

[2] Culver, D., "A Tangled Bundle?" *tele.com*, Vol. 3, No. 13, p. 51.

[3] "The Importance of Billing Earnestly," *America's Network Telecom Investor Supplement*, Vol. 103, No. 18, p. 6.

[4] Morri, A., "Carriers Lament Customer Confusion," *Telephony*, Vol. 233, No. 3, p. 50.

[5] Levine, S., "Q: When Is a Bill Not Just a Bill?" *America's Network*, Vol. 103, No. 14.

[6] Levine, S., "The Case for Self-Serve Care," *America's Network*, Vol. 103, No. 1, p. 80.

[7] Raaen, D., and M. Wolfe, "CRM Takes the Driving Seat for Shareholder Value," *Telecommunications, International Edition*, Vol. 33, No. 8.

[8] Meyers, J., "The Software Elite," *Telephony*, Vol. 237, No. 3, p. 24.

16

Technology Management

There is no longer such a thing as long distance.

—Chairman, global service provider

Among the many strategic decisions that all service providers need to confront is whether to take a leadership position in network technology. Of the three value disciplines of operational excellence, customer intimacy, and product leadership, this is the product leadership dimension. Contrary to conventional wisdom, a decision to excel in product leadership does not necessarily require huge investments in multinational networks. A service provider can sustain product leadership by controlling the user interface with innovation and reselling the majority of its physical infrastructure wherever clear distinctions do not exist among suppliers.

Other than Sprint, with its "pin-drop" campaign, few service providers brand their products with technology leadership. AT&T, long associated with technology leadership, made overtures to branding its technology prior to the Lucent spin-off, but its present emphasis in both wireless and consumer markets is on distribution rather than technology leadership. Qwest has tried to differentiate its IP network as a superior technology, but its limited reach to small customers will probably prevent it from developing its brand before better-known service providers have their own IP infrastructures. It is likely, though, whether it is Qwest or another provider, that an IP-based service provider will take a product leadership position in the industry.

16.1 Innovation and disruptive technologies

In the book *The Innovator's Dilemma* [1], Christensen introduces the concepts of sustaining and disruptive technologies. A *sustaining* technology is a technological advance that improves the performance and profitability of existing products. DWDM, which improves the cost-effectiveness and bandwidth of fiber, is a sustaining technology. Customers of existing technologies tend to migrate to the improved benefits of sustaining technologies. *Disruptive* technologies do not necessarily meet the criteria of faster, better, or cheaper, but they do offer a unique value proposition to customers that are not satisfied by present technologies. The first mobile phone was harder to use, more expensive, and lower quality than its wireline equivalent, but it created a new group of customers for whom an additional landline did not supply value. Services based on disruptive technologies do not appeal to the present customer base; if they did, they would be sustaining technologies. Typically, disruptive technologies result in poor performance compared to existing technologies, but they offer a different set of benefits. Eventually, successful disruptive technologies tend to overcome quality shortcomings. Fax over the Internet and IP-based telephony are both disruptive technologies.

Because the markets for sustaining and disruptive technologies are essentially separate, it is not necessary for service providers to commit to one or another, if they can afford to include both. Investments in IP by AT&T, MCI WorldCom, and Sprint signify that these carriers are willing to maintain infrastructure duplication in anticipation of the development of the IP market. Abandoning sustaining technologies in favor of disruptive technologies is not advisable for incumbent service providers. Disruptive technologies generally offer lower margins than existing services. The markets for disruptive technologies are not in the mainstream and are often dramatically smaller than the service provider's existing customer base. Existing customers of sustaining technologies reject disruptive technologies, so a service provider's decision to switch could drive away its core customer base. On the other hand, neglecting disruptive technologies during their market development could position a service provider at a significant disadvantage when the new technology begins to compete on the same terms as the sustaining one. Ironically, listening too carefully to customers could cause a service provider to disregard the disruptive technologies until it is too late to deploy them successfully.

New entrants with clean investment slates and no brand image often have a market advantage over incumbent providers for whom low quality would tarnish their existing brands. Incumbents cannot risk using disruptive technologies for the small incremental increase in the customer base that they would provide, especially if the new technology has the potential to drive away loyal customers. New entrants are delighted to gain even a small fraction of an incumbent's customer base. Still, incumbents have successfully used anonymity to gain market share in niche markets. One example is the dial-around calling providers in the United States. Dial-around, also called casual calling, occurs when a customer dials a code to bypass the long-distance network of the presubscribed provider. In the United States, dial-around calls require the customer to dial 10-10-XXX. Billing for the calls appears on the local service provider's statement. The benefit of dial-around calling is low price, but the caller needs to dial extra digits to reach the dial-around provider, so the service is not as convenient as direct dial. The market for these services includes students and other cost-conscious customers and customers whose credit history excludes them from presubscription to a long-distance provider. While thousands of local entrepreneurs once offered dial-around service, the leading providers are now AT&T and MCI WorldCom, which have added to their markets and hedged customer loss by offering inexpensive dial-around. To avoid any impact on their existing brands, neither company provides any identification information on the many commercials and other branding for their services. Service providers that want to test disruptive technologies need to distance their existing businesses from the new organization in the same way.

One way that disruptive technologies enter the mainstream is through the pace of innovation. The original version of any technology is always below the customer's needs. The first version of an innovative software package, when it is a disruptive technology, does not contain all the features users need, partly because the software developer needs to ignore some features to get the software to the market, and partly because users imagine new features as they become familiar with the product. Later versions come closer to meeting user needs, and eventually the package becomes feature rich, with niche capabilities for only a small segment of the user base. It is then a sustaining technology and improves at a slower pace than it did while it was still a disruptive technology. Customer needs grow at a still slower rate than these late product enhancements, and eventually the software is capable beyond the capability of most users. If a disruptive technology emerged during this development process and improved at approximately the same rate as the sustaining software, eventually the disruptive software would meet the needs of all but the most advanced users. If this technology were also less expensive or had some new benefit not available to existing users, the mainstream customer base would be most interested in converting, if conversion is reasonably convenient.

IP telephony is presently in the disruptive position to traditional circuit-switched telecommunications. It is still below the quality of traditional communications, and it has not demonstrated a clear enough price advantage to migrate mainstream customers. IP voice telephony demonstrates many of the qualities of a disruptive technology. The first voice-over-IP systems required users to own PCs equipped with a sound card. Anyone the user called needed to have a PC and the same proprietary software. Transmission quality was barely acceptable. As the technology progressed, the transmission quality began to approach that of wireless service and service providers established gateways to eliminate the need for users to have computers. Like most disruptive technologies, IP telephony moved from a niche status to a threat to established providers and eventually to a mainstream technology. In the near future, further advances in IP telephony will be sustaining instead of disruptive technologies.

Still, IP telephony promises to add value that is barely imaginable by the present user of telecommunications services. For one thing, billing options become virtually unlimited in an IP platform. Time and distance matter very little, and both the amount of bandwidth used and the priority of the transmission are manageable by the service provider or the customer. For customers that are interested in usage-based billing, IP telephony is quite capable of accomplishing their objectives. Carriers will be able to create tiers of service quality and price them accordingly, as compared to the present practice of price discrimination that creates an arbitrage opportunity in a competitive market. Insight Research estimates that the IP billing market will grow from $110 million in 1998 to $5 billion in 2004.

IP holds the promise of capabilities that exceed those of the traditional network, once the quality barriers disappear. IP emphasizes logical over physical connections, facilitating network sharing by multiple service providers. IP can include familiar circuit-switched network features such as call waiting and conferencing and is likely to be superior in areas that have not been available or successful for incumbents: videoconferencing, whiteboarding, and video attachments. Unified messaging is gaining support among computer users, and IP telephony is an excellent if not superior platform for this new capability.

New entrants need to select a technology upon which to focus, and many are naturally attracted to the disruptive IP architecture. The decision facing incumbents with existing facilities is that of timing their investments into new technologies instead of continuing to invest in the sustaining technologies that comprise their existing businesses. Leading equipment manufacturers such as Lucent and Nortel are making that decision easier for the leading service providers by developing technologies to ease the migration from circuit-switched to packet-based networks.

Disruptive technologies can create a path to differentiation for new entrants and potentially for incumbents as well. New entrants whose goal is to serve a market segment rather than the universe of users can select a disruptive technology that sustains special benefits for its intended target market. Incumbents whose chief competitors or customers are quite risk-averse can take the risk of a disruptive technology, once an initial analysis demonstrates that the new technology will not erode the base. Incumbents that predict that the new technology will surpass the current sustaining technology can introduce new services through a new brand or new distribution network, protecting the existing base.

16.2 The network effect

The so-called network effect phenomenon exhibited by telecommunications networks has less to do with physical networks than it does with critical mass. The network effect occurs when the value of one customer's experience increases sharply, perhaps geometrically, as the number of customers grows. This effect is somewhat counterintuitive to conventional supply and demand, in which the value (or at least the cost) of a good should decline as production increases.

In the telecommunications industry, management and regulators did not name the network effect, but they recognized it and priced accordingly with subsidies. The telephone was most valuable if nearly everyone had a telephone. Therefore, regulators concluded, pricing for telephone service should amortize the high first cost (a barrier for many customers) and keep monthly rates low. Business users paid higher rates than consumers did because it was in their own best interest to get their customers connected to the network. Thus, the last-mile challenge perplexing regulators and providers today stems indirectly from the value of the network effect, recognized a century ago.

The Internet has illuminated the network effect for many Web-based businesses. One reason that auction house eBay draws customers is that new customers assume that many vendors and auction items are already there. Customers are less likely to patronize auctioneers that they do not think are popular (although one could argue that the right item at an unknown auction site would sell for less if fewer purchasers are bidding). Growth of the Internet itself was a product of the network effect. When only hobbyists were using the Internet routinely, corporations and other businesses could not gain enough value from Internet surfers to invest in their own sites. Once a critical mass evolved, most businesses recognized the opportunity to profit from Internet investment. As businesses and other institutions placed information and transactions on-line, pages became user-friendlier and the Internet became more attractive to new users. While only a few years ago most mainstream corporations did not maintain a home page, e-commerce has become an information systems investment focus. Similarly, early adopters with Internet access were less willing to go on-line when the on-line content was paltry, and there were so few other e-mail users that mail was an unusual occurrence.

Services that demonstrate the network effect can often benefit from first-mover advantage if they use the opportunity to gain critical mass. One advantage of the network effect is that it occurs whether or not the service provider owns proprietary technology or maintains a network standard. AOL enjoys the network effect; one reason consumers subscribe to AOL is that they believe everyone else does. Other than its renowned instant messaging capabilities (which are primarily of interest to children, not adults), AOL boasts few real technological advantages. Customers of any ISP can reach the Internet, send e-mail to AOL subscribers, and use many AOL capabilities from its home page. Standards do not draw new customers to AOL with the same attraction as the network effect does. Service providers enjoying the network effect find that their customers are more loyal and that acquiring new customers is less difficult.

The network effect creates a sort of chicken-and-egg dilemma. Service providers that want to invest in a new service need to time their investments and account for the growth in critical mass necessary before the service is profitable. The service provider cannot delay the investment, because first-mover advantage is critical. The service provider must be able to survive without profitability in the service line until critical mass occurs. Last, the service provider must constantly enhance the service with new features to ensure that competitors cannot attract the customer base once there are sufficient customers for profitability.

For basic Internet access, the network effect will probably have a significant impact on price that conforms more to supply and demand. As the number of Internet customers rises, the value increases to the customers, and the value of the pages increases to advertisers. ISPs have historically needed to cover their costs through subscription fees and augmented their revenues through advertising. As the number of viewers grows, the base becomes more valuable to advertisers, and ISPs are already beginning to eliminate their subscription fees. Free access will undoubtedly create a larger user base, which will raise advertising revenues, which will then support new incentives to view pages, and the cycle will continue.

The network effect can create a path to differentiation for customers who want access to other customers or to content from the most popular source. For service providers that plan to launch a portal or a specialized information service, the strategic planning process needs to consider the network effect and the ways it can promote the success of the new service.

16.3 Research and development strategies

Competition has transformed the R&D model for service providers. Monopolists can control the nature and pace of product development, but competitors respond to market needs. Before deregulation (and before the technology explosion), a monopolist could vertically integrate to include an in-house R&D function. Many of the leading equipment manufacturers such as Lucent and Nortel were rooted in the vertical integration of a monopoly provider. While some R&D will undoubtedly move back into the service provider's organization, the trend to outsourcing will probably continue. For one thing, the development cycle is only about six months, and a service provider with other priorities simply cannot devote the attention it requires to service creation as well. Service providers can maintain relationships with their R&D partners and arrange to develop proprietary technologies, which provides nearly the same benefits to the service provider as in-house development would.

Telefónica's in-house R&D group resolves the time-to-market challenge by building its applications from a core technology it uses as a software platform. The organization adapts the platform to company-specific requirements through development agreements with its partners in Europe and South America. Its development investments have grown 10% per year since before 1990.

AT&T Wireless has maintained a targeted in-house R&D facility [2]. Competitive local provider Nextlink maintains a working central office and network operations center as a laboratory test facility. By testing planned equipment upgrades in the laboratory, the service provider is able to reduce the deployment problems they would face in the field. The CLEC also provides feedback to its vendors about their products' performance in its own operating environment. Sprint PCS maintains its Systems Technology Integration Center near the company's network operations center. Its commitment to a multivendor infrastructure necessitates testing the interactions between its vendors' equipment. CLEC ICG launched VoIP service in hundreds of cities. The service provider uses its R&D laboratory to track IP telephony technology, including unified messaging, conferencing, and other features. Long-distance provider IXC Communications (now part of Broadwing) maintains two development environments, one that mimics its fiber optic network and one that tests new equipment.

The constant reminder of competitive intensity has encouraged service providers to focus more on practical applications and less on heuristic development. AT&T's prize-winning research was often of great benefit to the scientific community but not as fruitful to its own bottom line. AT&T's research budget is now 10% research and 90% product development [3]. France Télécom's research facility boasts seven targeted units: four directed toward services, two to network integration, and one to network management. The company also shares a U.S.-based research investment with Deutsche Telekom. EURESCOM, a research venture backed by 24 service providers, allocates research projects for the benefit of its members.

Many service providers are also teaming with their equipment suppliers or universities to develop research. Besides pooling funds and skills, the advantage of teaming vertically on the distribution chain is that changes to working relationships through mergers and business expansions do not generally threaten the partnerships. For example, France Télécom and

Deutsche Telekom forged their research venture while their partnership in Global One was still active. Telecom Finland decentralized its R&D operations to the business units, where teams work alongside customers [4]. Its chief competitor, Finnet, has taken the opposite approach; the consortium of local providers carries out joint R&D.

R&D is a critical element for those service providers that choose to differentiate their services based on service quality. These service providers will need to decide as part of their overall strategy what percentage of service development is conducted in-house and how to retain proprietary rights to research performed by third parties on its behalf. All providers, whether or not they plan to differentiate through network quality, will need to monitor developments in the technologies supporting their businesses.

R&D investment is one area in which the largest providers enjoy a significant advantage—if their smaller competitors try to compete on the same terms. Many service providers use a percentage of revenues as a rule of thumb for budgeting R&D. A smaller competitor that is trying to catch up an AT&T technology lead will find the gap widening under even the best of circumstances. The R&D budget at AT&T is undoubtedly as large as the entire revenue stream of many of its competitors. All other things being equal, AT&T's investment will widen the technology gap, whether AT&T starts out slightly ahead or far ahead in the development race. The only way that small competitors can hope to compete against AT&T's service development is to focus their R&D resources in a specific area—to make their apparent investment in the area of focus larger than AT&T's dispersed investment in the same area. In addition, competitors need to ensure that the area of focus is not so strategically significant to its larger competitor that it will retaliate with additional development expenditure.

16.4 Service creation

Deregulation of the telecommunications market has clarified for service providers what has always been clear to competitors in other industries: Customers buy solutions, not services. As dial tone becomes more in demand, it has also become more of a commodity, and few service providers view conventional access and transport as the service of the future. Furthermore, the convergence of computers and telecommunications has occurred in the network as well as on the desktop, and service providers are freer to create proprietary, software-based services that can differentiate their basic commodity services.

Equipment manufacturers are developing a new class of switch called softswitches, which use programming intelligence to integrate voice and multimedia services over telecommunications networks. Softswitches separate the call control functions from the transmission function. Service providers can imagine, program, and launch new services without the intervention of the vendor, moving from strategic planning to market implementation almost instantly. Softswitch platforms can enable customers to configure their own profiles, add new services, and activate them instantly by using a familiar browser interface. Besides reducing the cost of provisioning for the service provider, this feature adds to the user's convenience and enhances the service provider's brand. As switching standards evolve, third-party developers will undoubtedly create an applications market potentially as active and diverse as the PC software market. Moreover, because softswitches cost much less than comparable conventional switches, and because of their scalability, new entrants will find facilities-based market entry to be affordable.

Successful deployment of new service creation methods will require vision and investment from service providers. Information systems such as OSS and financial systems need integration and must be available through proper security to network-based applications. Service providers will need to design simple interfaces that clearly brand their applications. Many service providers will commit dedicated resources to listening to customers so that they can launch new applications before customers forget that they requested them.

For an incumbent, superior service creation can transform its image from a lumbering and undifferentiated utility to a flexible, customer-focused brand. For new entrants, service creation can establish a targeted niche business model or compete head-on with large incumbents that have failed to make a service creation commitment.

16.5 Budgeting for technology expenditures

Accountants classify most technology expenditures as capital investments, which means that the service provider will gain a benefit over more than one accounting period. Companies can expense (rather than depreciate) certain investments that have low per-unit cost, even though taken together they will probably last for a longer period and can indeed represent a large expenditure. For strategic planning purposes, it is worthwhile to include in the budget process all technology expenditures that will last more than a year regardless of the way the accounting system will handle them.

The capital budgeting process attempts to balance the service provider's capital resources against improved shareholder value and strategic objectives. If resources are unlimited, the task is to decide which technology investments will provide an acceptable rate of return and initiate all of them. More often, resources are limited, and the job of technology planners is to decide among competing investments. The analysis needs to consider the potential return from the new technology as compared to alternative investments, the risks associated with the return, strategic objectives of the service provider, and the impact of the investment on other business processes.

The first step in technology capital management is sizing the capital program. This process requires consideration of long- and short-term objectives, nondiscretionary technology initiatives such as local number portability in areas undergoing deregulation, and the amount of capital available. The capital available to discretionary projects is the remainder of the budget, after the service provider funds nondiscretionary projects. Many service providers use both a top-down and bottom-up approach. The top-down approach starts with a total investment amount, developed from previous capital programs or a competitive assessment. The bottom-up approach simply adds together the capital requirements of all the desired technology initiatives for the planning period. The next step is to align the expenditures to strategic and operational objectives. Tests for evaluating individual programs can include financial measurements such as NPV or EVA. One significant change to these analyses is the dramatic reduction in the expected life cycle of hardware. Switches, which once commanded depreciation schedules of 20 years, now last only a few years before they become obsolete. Risk analysis of individual projects is also part of this phase. When a proposed technology initiative does not meet hurdle rates, it is appropriate to evaluate its nonfinancial benefits such as its strategic importance, competitive parity, or customer demand. The third step is to include the required expenditure in the capital budget. During this step, management selects performance measures to ensure that the expected benefits occur and assigns accountability for overseeing the initiative. Operational performance measures can include asset utilization, revenue enhancement, and response time. The next task is to monitor the implementation of the technology program. Besides ensuring that the project launch is successful, this phase can provide useful information to future planners about the accuracy of their projections. Too often, planners estimating the costs and benefits of technology initiatives use historical projections instead of actual historical data. When planners test their own assumptions and methods against the actual implementation, they can improve their own forecasting techniques for the next analysis. The last task is to redeploy or dispose of underperforming facilities.

16.6 Managing service introductions

Telecommunications service providers have always experienced chicken-and-egg dilemmas related to service introductions. Most new services have very high fixed costs. Even the central office-based enhanced services such as call waiting and conferencing had large deployment costs and negligible variable costs, because each central office offering those services required an upgrade against otherwise usable equipment. Furthermore, like other disruptive technologies, these services were difficult to price. Customers could not imagine what they would be willing to pay; they could barely imagine how they would use the new services. Consequently, telecommunications service providers often had less information than they would have preferred to help them make deployment decisions.

Some technologies also suffer from switching costs, because customers have invested in older technologies. DVD is an example of a technology that needed to eliminate or mitigate switching costs. One medium it replaces is CD-ROM, ubiquitous in PCs. Users contemplating DVD for data storage would not have embraced the technology, considering their large stores of CD-ROM software and file storage. DVD eliminated this problem by making the devices compatible with legacy CD-ROMs. Another problem facing technologies is the chicken-and-egg-problem. DVD as a substitute for VCRs cannot be highly competitive until many software titles are available. The chicken-and-egg problem is that distributors are unwilling to produce millions of copies until there is a critical mass of buyers, and the per-copy cost of software is prohibitive if the manufacturing costs cannot be shared among many customers. People will not migrate to DVD until many titles are available; titles will not be available until a sufficient number of users have made the commitment to DVD.

This is another example of the way the network effect reduces prices and improves value to customers. Telecommunications service providers have traditionally upgraded to new technologies quite seamlessly, but part of the reason for that was the partnerships between dominant providers and regulators. In a highly competitive global market, these more jagged migrations will be more frequent.

Telecommunications service providers will need to make decisions that will affect service introductions: Should the first service have a full set of features or a simple design that will attract customers? How can we manage the price of the new service when early adopters will pay a premium, but there are significant strategic advantages to gaining a large critical mass of customers as soon as possible? Is it appropriate to offer the service well below its fully allocated cost, knowing that a large mass of customers will reduce the costs and that technology advancement will undoubtedly reduce the cost as well? What is the impact of this service on revenues from other services?

A service targeted to early adopters will gain a price premium, but probably at the expense of a relatively small market characterized by churn as soon as a newer service emerges. Targeting the new service to a wider base of customers requires additional investment in customer service support, user-friendly materials, yet the mainstream customer is less willing than the early adopter to pay a premium for new services compared to the existing portfolio. For services directed at the low end of the market, separate brands and separate distribution systems will limit the impact of the mass-market services on the principal brands.

16.7 Supporting tools

Planning networks has required computer-based tools since they became available. Proprietary network planning tools, on their own, could enable service providers to differentiate their networks to their customers. Service providers who choose to operate in the wholesale channel will find that sophisticated network management tools will be useful for internal operations and for marketing to technologically advanced customers.

Network management and design tools

Many tools are available to assist telecommunications service providers in optimizing their network operations, both for their ongoing operations and in network design. Because most of the design and operations tools are proprietary to the technology they support, they differ by manufacturer and application. Customers, too, can use network management tools to monitor the service they receive from their service providers and manage their service portfolio. Often these tools are required as part of the service-level agreement between the service provider and the customer. Besides providing operational information to ensure the reliability of the network, they can produce planning data to use in capacity planning, market planning, and service creation.

Technology management process tools

Decision support tools are available to assist in the budgeting process, and managers will find that statistical tools will help to estimate the impact of the network effect or the adoption of new technology and its impact on profitability. Furthermore, total quality management programs and business process reengineering support dozens of tools that can assist strategic planners in optimizing their technology management efforts.

Some service creation tools have been available for several years but are not yet in wide use. Design techniques such as features analysis or molecular modeling can evaluate the mix of services to offer to customers [5], and process blueprinting charts the service process.

16.8 Role of information technology

Information technology is integral to the path of technology management to differentiation. In some ways, this entire chapter describes the influence and importance of technology upon a service provider's ability to differentiate its offerings to customers. The switching and transmission network is moving toward a software-driven architecture that will make it virtually indistinguishable from other information applications within only a few years. Some industry observers have made the argument that IP telephony is simply the application of computing to the telecommunications network. Furthermore, all of the tools and supporting infrastructure, such as OSS technology, network management, and service creation, all of which used to reside within the switch, are now separate manageable information systems applications, ready to integrate with enterprise resource management systems and other corporate applications. Therefore, network technology management is virtually inseparable from information systems management.

References

[1] Christensen, C. M., *The Innovator's Dilemma*, Cambridge, MA: Harvard Business School Press, 1997.

[2] "Proving Grounds," *Telephony*, Vol. 236, No. 13, p. 20.

[3] McClure, B., "Less 'R' and More 'D,' " *Telecommunications*, Vol. 33, No. 7, p. 73.

[4] Evagora, A., "The Nordic Track," *tele.com*, Vol. 3, No. 1, pp. 68–72, 75.

[5] Young, L., "Communicating in a Consumer's World," *Telecommunications, International Edition*, Vol. 33, No. 8.

17

Business Expansion

It's niche, merge, or die.

—U.S. research analyst

17.1 Increasing market share

In a growing market, maintaining market share will increase revenues at the same pace as the market is growing. In a stagnant or declining market, increasing revenues requires increasing market share. A telecommunications service provider can increase market share in three ways, and these measures are not mutually exclusive. First, the service provider can take customers from its competitors. Second, the service provider can reduce its own churn to below the industry average, thereby raising its share against its competitors. Third, the service provider can buy more market share through mergers and acquisitions. This is essentially the technique used to great advantage by WorldCom and is discussed in Section 17.5.

Each method of increasing share holds its own costs and risks, and the prudent service provider is constantly seeking opportunities through all feasible opportunities. The service provider can take customers from its competitors by using a variety of techniques, including promotions, price wars, and brand differentiation.

Telecommunications service providers commonly use *promotions* to attract new customers. In fact, promotions are so common that customers seeking to switch providers often demand that the new service provider offer some incentive to join. Therefore, promotions have several disadvantages.

- *Promotions attract undesirable customers.* The most loyal customers, by definition, are difficult to acquire from a competitor. Similarly, customers that are easiest to acquire are unfortunately always receptive to new offers. Therefore, the customers that seek out promotional pricing will continue to seek it from competitors no matter whose customer they are temporarily. In addition, customers that respond well to promotional pricing are often very price-sensitive. This creates a commodity environment, trimming margins and encouraging churn.

- *Promotions create a prisoner's dilemma.* Promotions only occur within a visible marketplace where competitors see—and potentially retaliate against—every competitive move. Therefore, taking a first-mover position with promotions can create a prisoner's dilemma (as described in Chapter 11) in which the major competitors have no choice but to match the promotional offer. In that situation, every service provider ends up with an equal and spiraling cost of customer acquisition and no provider sustains a competitive advantage.

- *Promotions set up a churn mentality when the promotion period is over.* Promotions, by design, are temporary. A long-distance provider could offer a tempting number of free minutes for the first month or several months of service. While this offer creates a switching cost for the duration of the promotion, it also produces a startling realization for the customer when the promotion is over. Wireless providers often find that their customers are most likely to churn at the end of the required contract period. Promotions often act as a soft contract, and customers are ready to churn as soon as the promotional period is over. This is a reason not to use promotions as a loss leader, because many customers will be happy to remain during the loss period and switch before their life cycle would be profitable.

- *Promotions annoy loyal customers that do not qualify for promotional pricing.* Promotional pricing for new customers runs the risk of displeasing existing customers, who properly argue that they have earned the promotional pricing (or awards) by their continued loyalty. Service providers certainly do not wish to gain promotion-seeking opportunists at the expense of the loss of loyal stalwarts. Some telecommunications service providers have minimized the visibility of their promotional offers by limiting them to one-on-one telemarketing to new subscribers rather than including their details in national advertising that existing subscribers would certainly view.

While there are many challenges to using promotions, promotions are invaluable tools in some situations. Promotions are exceptionally effective for migrating existing customers to new services. Customers who do not know how they would use a new service such as unified messaging would appreciate a trial period, especially at no cost. Those who decide that they like the service can keep it for its normal fee, and these customers are unlikely to shop elsewhere, especially if they believe that there are switching costs. Switching costs for unified messaging might include addressing, an interface with an e-mail client, setting up a profile, or becoming familiar with the interface. Any of these, and certainly all combined, can serve as switching costs to the customer, especially if competitive alternatives offer no clear advantages. Promotions can be effective for attracting a new customer segment, such as low-usage wireless customers, assuming that other service providers offer no apparent price or quality differentiation once the promotional period is complete. The promotion can act as the motivation to subscribe to a service that is otherwise not a priority for the customer.

Price wars are often the last resort of the commodity seller. Price wars tend to benefit only the customer, for many of the same reasons as promotions. Competitors in intensely competitive markets and commodity markets react to price reductions with their own price reductions. When a price reduction occurs outside of a cost reduction, only the service provider's margins drop. On the other hand, when demand is elastic, as it is in many telecommunications markets, price reductions do raise usage, and the improved asset utilization can create higher profits. Therefore, service providers need to use sophisticated economic modeling techniques with solid assumptions before embarking on a price war. While there can be economic benefits to engaging in a price war, price wars do not guarantee an increase in market share.

Service providers also use *brand differentiation* to draw customers away from their competitors. Branding a service requires that customers perceive it as differentiated from its competitors. Sprint has successfully branded its network and wireless services based on quality. AT&T's brand endures partly because of its longevity as the incumbent monopoly provider and partly due to its more recent efforts in the areas of innovative pricing. Incumbent local service providers are investing considerable expenditures in name recognition and branding to ward off potential and actual competitors and develop loyalty. Branding is mysterious; after its acquisition by WorldCom, MCI dropped from second to eleventh in aided brand awareness in 1999, according to market researcher IDC. The same survey identified Microsoft, Nokia, and Sony in the top five telecommunications service providers, though other than Microsoft's MSN and WebTV ISPs, none of these is a service provider at all. One interpretation is that these three industry suppliers would have an easier path than others, should they choose to integrate vertically into telecommunications services. An alternative inference suggests that brand awareness is elusive, especially in light of the significant service provider investment in image advertising. Branding works to increase market share because it creates an uneven marketplace. A service provider with a valuable brand can command higher prices for services identical to those offered by its competitors, a higher market share than an unbranded equivalent service, or both. The challenge to branding is that the service provider can invest in its branding campaign, but the brand itself exists only in the mind of the customer. Advertising, network quality, and excellent customer service can help to build the brand, but in the end, the control of a brand is primarily indirect for the service provider.

Churn reduction is another technique for gaining market share. As long as one service provider's churn is less than that of its competitors, the service provider will gain share, all other things being equal. A recent study by IDC determined that respondents were more likely to change their long-distance provider than any other telecommunications service. Nearly one-quarter of respondents had switched long-distance providers in the last year. The most common reason for the churn was price. For Internet users, the most common reason for churn was poor service quality. In Europe, the Internet churn rate has increased to about 25%, according to the Strategis Group. Andersen Consulting estimates that wireless churn in Europe is also 25% [1]. U.S. wireless churn increased from 24% to an estimated 33.6%, when an increase in the number of service providers created competitive pressure to eliminate the service contract [2]. In the United States, the Strategis Group also reported that about 36% of wireless users intend to switch service providers within a year, most often for price but also to obtain number portability. Unfortunately, the customers most likely to churn had an average monthly wireless bill that was 37% higher than the bills of the nonchurning respondents. Therefore, reducing churn not only increases the customer base, it increases the average revenue per customer. The survey also provided some insight into the phenomenon of churn itself. The Strategis Group survey did not report actual churn rates; it described instead the customer's intention to churn. As number portability was not available at the time of the survey, it also provides insight into the attractiveness of churning itself to the customer. Customers will churn simply to get number portability. The chief benefit of number portability is that it makes future churning more convenient. Therefore, customers apparently value the ability to churn outside of other service characteristics. One method for reducing churn is bundling services. Customers who have multiple services from the same provider are less likely to churn. An Alexander Group survey of long-distance providers concluded that customers with a service bundle that includes local services are one-eighth as likely to churn as other customers [3].

Service providers can draw several conclusions from what they already know about churn. First, customers value the right to churn, so services that prevent customers from leaving will have a disadvantage over services with low or no penalties for terminating service. Second, customers churn more often in services that are undifferentiated other than price (including, for example, the price of a wireless handset). Third, eliminating only one-half of existing churn can provide an increase in market share that rivals the performance improvement of virtually any technique.

17.2 Increasing market size

Another method to create growth is to increase the overall size of the market, keeping market share constant. Often, the market leader will create image advertising that intends to increase the overall market rather than its own brand within the market. AT&T's "Reach Out and Touch Someone" campaign lasted longer than its monopoly. AT&T used very emotional appeals such as the personal relationship between sisters living miles apart or remembrances of special moments with a parent to convince customers to make purely noncommercial calls. The campaign barely competed against competitors (indeed, for much of its duration, there were no real long-distance competitors). Rather, AT&T's "Reach Out and Touch Someone" campaign competed primarily against the postal service. After competition began, AT&T recognized that increasing the market size increased its revenues and was a more effective strategy than attacking its competitors would be. When AT&T enjoyed 90% market share, it could expect that listeners to the campaign who decided to make long-distance calls were much more likely to use AT&T than its competitors. Recognizing the presence of competitors, AT&T made certain that its own brand was visible in all its advertising, but the focus of the campaign was to increase the long-distance market.

In a highly competitive telecommunications market, it is nearly impossible to expect that any provider will have 90% of a mainstream market, but it is plausible that one provider could enjoy a large majority of a market niche. Campaigns to increase the size of the market—by emphasizing the benefits of the service, not the differences of the individual provider—will be effective in achieving corporate growth for the market leader. MCI tried something similar with its "Is this a great time, or what? :)" campaign for data services. Furthermore, when competitors are generally offering undifferentiated services, a campaign to sell a generic service can be more effective in increasing growth than a campaign to differentiate the provider might be. For service providers that are not the market leader, and leading providers with less than about two-thirds of the market, management will need to compare the cost of an image campaign against the incremental revenues that the service provider can expect to gain.

Service providers can increase market size by adding new customers, either through territorial expansion, or through finding new market segments to serve. Territorial expansion includes the efforts by RBOCs and new local service providers to create a presence in additional cities. SBC's commitment to compete with incumbent local service providers in 30 U.S. cities outside of its existing coverage will somewhat ironically increase its market size as required by regulators to temper its large merger. New entrants often launch their services in the most lucrative markets (generally the largest metropolitan areas) and then expand into second and third tier markets to increase their base.

Territorial expansion also includes the globalization initiatives by the world's leading providers. The largest service providers will offer worldwide services, not only to increase their potential customer base. A global presence enables the service provider to offer a valuable service package to multinational corporations, a very attractive customer segment. Multinationals often find that they need to maintain relationships with multiple service providers and a mosaic of price structures, services, and customer relationships. A global service provider can eliminate much of this customer inconvenience, provide standardized service offerings, and reduce the customer's cost of an in-house telecommunications staff.

17.3 Vertical and horizontal integration

Service providers can grow through vertical and horizontal integration. While this is a relatively easy strategy to execute, succeeding in either of these endeavors is quite challenging. Vertical integration is a service provider's migration along the distribution chain. When a service provider buys a supplier or a distributor of its services, the resulting transaction represents vertical integration. The effect of vertical integration is that it widens the service provider's presence on the distribution continuum. Global Crossing's purchase of Frontier and Qwest's purchase of US West each represents strategic decisions on the part of the wholesale providers to acquire retail distribution. These are vertical integration moves. Chapter 12 discusses the benefits and risks of vertical integration.

The mergers between Bell Atlantic and NYNEX and GTE and SBC's acquisitions of Pacific Telesis, Ameritech, and SNET represent horizontal integration to gain additional territory. The same is true of Vodafone's acquisition of Mannesmann and each of these wireless providers' earlier initiatives (such as Mannesmann's acquisition of U.K. wireless provider Orange and Vodafone's acquisition of Airtouch). The acquisition of MCI and the attempt to acquire Sprint by WorldCom represent horizontal integration to broaden the customer base. AT&T's buildout through the acquisitions of wireless carriers were horizontal integration initiatives to broaden the customer base and acquire network infrastructure. Horizontal integration expands a service provider's product lines within its existing distribution chain. Companies integrate horizontally for a variety of reasons. In the telecommunications market, one major justification for acquisitions is to increase the scale of the company. Acquisitions to achieve scale must achieve significant cost reductions in the duplicate administrative structures of the merged company. Companies with very different infrastructures or cultures will find this a challenge, intensified by the likely presence of an acquisition premium paid by the acquirer. It is not sufficient to eliminate duplication; somewhere new profits that cover the premium need to emerge, without any obvious new business opportunities.

Strategic alliances create some of the benefits of outright merger without many of the risks. AT&T and British Telecom established their Concert alliance to serve a nearly global market while avoiding the learning process and considerable investment that would be required to do so alone. The structure of the alliance is more durable than a simple marketing agreement; each partner has invested billions of dollars and transferred international facilities to ensure its success. The now-defunct Global One intended to accomplish the same goal, but the announced acquisition of Sprint disrupted its plans. A similar disruption ended the alliance between British Telecom and MCI before the Concert transaction. Bell Atlantic's premerger alliance with NYNEX gained some of the benefits of a large wireless footprint. After AT&T's resounding success with its own nationwide plan and other wireless carriers' replication of the offer, SBC and BellSouth launched their own alliance for wireless service. Similarly, Vodafone Airtouch and Verizon, have launched an international wireless venture, undoubtedly encouraged by predecessor Bell Atlantic's earlier success. Alliances create apparent if not actual growth and intend to gain some of the synergies and critical mass of a telecommunications leader. Their advantage is that the partners are free to retain any level of independence they desire. Like all temporary partnerships, the lack of outright commitment prevents the partners from sharing private information, and each partner acts in its own interests, knowing that the partnership could end abruptly. Participants in partnerships are generally less willing than merger partners would be to make significant internal sacrifices that would benefit the alliance to the detriment of the service provider on its own. Both Deutsche Telekom and Telefónica appear to have eased out of the alliances in which they participated in favor of lone global expansion.

Horizontal integration avoids the channel conflicts inherent in vertical integration, but it has its own challenges. Horizontal integration, by definition, adds an organization that is essentially in the service provider's same business. This implies that about half of the entire management and administrative structure, including all business processes and information systems, is unnecessary once the acquisition is complete. Nonetheless, most merger integrations cannot simply eliminate half the resources and move on from there. Merger integration is a complex process. At best, the service providers have budgeted both time and resources to ensuring a successful integration. At worst, the inability to integrate a merger can either undermine the benefits anticipated from the merger or derail the merger completely.

Value innovation is another method to increase the size of a service provider's market. It differs from horizontal integration because it makes a nonintuitive combination, not the obvious consolidation. There are numerous emerging examples of this potential in the telecommunications market. Wireless service providers have entered into agreements with information providers to send brief information messages to wireless users on their cell phones. New devices allow customers to browse the Internet from cell phones or network-enabled personal organizers. Speculation abounds about new applications, from the Web-aware refrigerator that orders milk at the proper time to the cell phone that notifies a traveler of a discount for fast food in the visited city. The merger between AOL and Time Warner will undoubtedly introduce value innovation and expand the markets for both service providers, but service providers will not require mergers to benefit from alliances with information providers, hardware manufacturers, retailers, and others to create value innovation.

17.4 Sizing new markets

Before a service provider can estimate its own revenue stream from a new market, it first needs to size the market potential. While the process involves several assumptions, there are sources available that can help to develop sound estimates. For any market that has generated real interest in the business community, there are many sources of market forecasts, many of which are publicly available. Market researchers track markets that already exist, such as the U.S. long-distance market, and they can provide actual and forecasted data. For very general information, the data is available in press releases, interviews, and the trade press. For a new venture, though, the service provider often needs data in more detail than is available to the public. The same market research firms that are the market source for industry trade magazines often publish studies of thousands of pages (and for thousands of dollars). The studies frequently break down the summary data into regional distributions, market segments, and other categories that will be very useful to the service provider undertaking strategic planning for new markets. If the large studies are still not detailed enough or focused on the market in question, the same market researchers can perform custom research (ordinarily at a higher price). Besides meeting the service provider's need for very uncommon data, the resulting data is most often proprietary to the client requesting and funding the study. Therefore, the market data can provide competitive advantage to the service provider that is willing to make the investment.

Another advantage of market research data is that these firms often estimate markets that do not yet exist or are very immature. Using sophisticated modeling techniques, these firms have valued Internet telephony, unified messaging, and various wireless information services before they were available to customers in the marketplace. While the data and projections require some flexibility and bracketing in the planning process, they are often the best data available and are certainly superior to projections offered by suppliers and others with self-interests potentially affecting their research.

Market researchers can often provide estimates of the expected market share of their clients against the overall market estimates. Service providers that intend to calculate their own views without the market research tools available to researchers can produce broad estimates for planning purposes. These estimates will not be as accurate as those produced by professional statisticians, but they can provide order of magnitude results for a strategic direction. A forecast of market size should include information from a variety of sources, including economic and demographic data, technology forecasts, assessments of the political environment, and pricing data.

One particular challenge to telecommunications service providers is sizing the market for services that customers do not currently use. Market researchers can draw on statistical techniques, mining historical data, and their own skills (a core competence) to develop projections for nonexistent or emerging markets. While their credibility is at stake if their estimates are far from accurate, market researchers do not bet the same level of investment that a service provider might in pursuit of a new business opportunity or a new customer market. For this reason, service providers often create pilot programs or incubators before they make a very significant financial commitment to new services or new technologies.

17.5 Locating acquisition candidates

Few industry observers are surprised when acquirers pay a premium to acquire service providers that offer synergy and profitability to a service provider's portfolio. Still, WorldCom's offer for Sprint raised eyebrows when the offer more than doubled the company's market capitalization, and the price was more than triple its earlier purchase price for MCI. Nevertheless, valuation of acquisitions is less about the performance of a candidate and more about the acquirer's expectations for the future. In the case of Sprint, access to the nationwide mobile network filled an otherwise serious gap in MCI WorldCom's service portfolio. Furthermore, mobile licenses constitute a barrier to entry; MCI WorldCom did not have an obvious opportunity to create a wireless startup. In addition, acquisitions create a substantial customer base much more quickly—albeit much more expensively—than building the business from within. With AT&T building its own size and the RBOCs poised to enter long-distance markets, MCI WorldCom recognized that it could easily fall from its second-place position to fifth or lower as RBOCs, global service providers, and new facilities-based network providers gained market prominence.

Note that WorldCom needed to shift its acquisition strategy from its original growth strategy. During the consolidation phase of long-distance competition, facilities-based carriers were struggling against the giants on one side and the resellers on the other. WorldCom was able to negotiate favorable prices and terms; its acquisition targets understood that if their own deal fell through, WorldCom could locate and acquire another service provider. Sprint held unique advantages in its market share, its brand, and its wireless properties. Service providers looking for growth through acquisition will take a path that resembles the early WorldCom, rather than targeting a market leader. In any industry undergoing consolidation, there are many small to mid-sized enterprises that are underperforming, either because their size does not afford any economies of scale or scope, or simply because the industry is too crowded. Service providers that maintain a bias toward growth through acquisition are constantly searching for business opportunities and constantly evaluating the potential value of acquisition candidates. Furthermore, like WorldCom, the service provider that plans to grow through acquisition needs to hone its in-house integration skills, to ensure that synergies and cost savings occur soon enough to minimize the impact on the company's operations and financial performance.

There are many sources for information about companies available for acquisition. First, financial services companies will assist in locating candidates, and often support in-house teams that follow the industry and find candidates as they arise. The largest service providers often maintain staff assigned to conduct competitive analysis and market monitoring. These analysts often uncover acquisition opportunities if management provides them with tools and guidance to do so. Last, customers, and especially new customers, can be a source of information. The sales force offers a fine window on the marketplace. New leads often come from customers who are dissatisfied with the performance of competitors. If the source of the poor performance is a service provider's unwillingness to invest in the relationship, the competitor could potentially be seeking an acquirer.

The synergy from a proposed combination often serves as either a financial justification or a sound bite for the merging CEOs to extol at the press announcement. Synergy can reduce costs, create revenues, and improve customer service. Companies can achieve synergy when they can share an activity or the infrastructure or transfer assets from one of its operations to another. If a facilities-based local service provider merges with a long-distance provider, the resulting network offers synergy by enabling the combined companies to share sales and distribution responsibilities. The synergy of the merger between AOL and Time Warner presumes that some of the Time Warner content will transfer to AOL and provide value not present when the two partners are individual entities.

17.6 Supporting tools

Most of the tools supporting business expansion are PC-based, because the fundamental basis of the planning process is financial and operational. To this end, spreadsheets and statistical functions that are not industry-specific can meet many of the strategic planning requirements to manage the growth of the business.

Market sizing analysis

Service providers can create broad estimates of market size for planning purposes, but any significant strategic decision should include rigorous estimates, probably from a third-party specialist. Still, a market sizing analysis performed in-house can help screen growth alternatives for a deeper and more expensive review. If possible, the initial market sizing should include top-down and bottom-up approaches. An analysis to determine the penetration of a new on-line product can estimate the revenues (or transactions or other valid measure) as a percentage of total transactions in current use. For example, a portal targeted to a service provider's small business segment might attract 5–10% of small business customers. The bottom-up approach to the same market adds together the various subsegments for a total estimate. In the small business example, the analysis might assume that none of the one-line customers will use the new service, that only 1% of customers with two to five lines (or fewer than 10 employees, or customers in the medical market) will participate, and so on. A data warehouse can often provide historical data, especially if the service provider had launched similar services in the past. The market sizing analysis needs to consider longitudinal changes (how usage will grow) to calculate when a break-even point might occur. Geographical data or demographic data can help to tune the estimates. Still, the foundations for the final projections will still always be estimates. It is always tempting to assume that more variables or more analysis eventually creates accurate forecasts, but that is rarely true when statisticians predict the size of markets and is unlikely to be true for a high-level in-house forecast. The service provider needs to take into account the cost and time expended in the analysis phase and act when more attention will not necessarily provide a commensurate value in the sizing data.

Valuation analysis

Valuations are best when they include facts and sound judgment. Several methods are in use, and the final valuation estimate is often a combination of the results from each of these techniques. The three most common approaches are the income approach, the asset approach, and the market approach. Traditional measures such as discounted cash flow and contemporary techniques, such as EVA, base the valuation decision primarily on income performance. This technique is especially applicable to service providers with a steady revenue stream in a relatively stable market. For facilities-based service providers that are asset-intensive, an asset approach to valuation can be effective. These techniques evaluate the book value of the asset base on the balance sheet and project a company value from that. In the telecommunications industry, several segments are indeed asset-intensive, such as the wholesale market. One flaw in relying too strongly upon this method is that the quality of the assets can vary significantly in an industry where assets become obsolete in a few years and sometimes months. Because asset value is an accounting device, an asset on the books with a high value can be worth considerably less when compared to its replacement value. Another approach gaining interest is the market approach. For companies already trading on public exchanges, the market capitalization (share price times shares outstanding) is the present value of the company. For a private enterprise, the market approach bases its value on the market capitalization of companies with similar characteristics, such as the service portfolio, the revenue stream, and operational measures.

A sound valuation needs to include financial measures such as cost of capital, discounted cash flow, the price itself, and other factors that are more difficult to quantify. These factors can include regulatory considerations if one or both entities are under regulation, currency exchange rates and other considerations if more than one country will participate in the new entity, synergies (business anticipated for the combined service provider that neither partner could obtain on its own), and management of the new entity. Costs of disposal for assets not needed by the new entity, the costs of integrating business processes, and the costs of eliminating duplication in the work force are all part of the valuation analysis. Valuation consultancy BIA has demonstrated that churn can reduce the market value of a telecommunications service provider. In one example, BIA establishes that a 0.3 percentage point rise in a typical wireless provider's churn rate reduced its market value by nearly 35%, considering its impact upon cash flows, required rate of return, and growth [4]. Financial considerations related to the structure of the transaction such as loan covenants or buyout considerations or pooling of interest accounting can all affect the soundness of the combination. Therefore, the team evaluating a potential acquisition needs to include skills in financial analysis, strategy, the legal and regulatory environment, operations, and negotiations. The prudent service provider engages outside specialists to assist in the valuation process.

17.7 Role of information technology

For the most part, business expansion is a series of decisions, so PC applications are the chief planning technology. Besides the obvious value of spreadsheet programs, statistical tools are available to assist in the market planning process. Most of these tools require some knowledge of business statistics, but the tools themselves have become more user-friendly than their mainframe-based predecessors. The other advantage of technology in fostering business expansion is in the exchange and common editing of information. Teams assigned to acquisition screening, due diligence, and integration need to exchange information regularly, and most of these efforts have very short deadlines. For alliances and possible mergers, the information exchange crosses company boundaries. Technology has facilitated much of the merger activity of the last few years and may have played a significant role in making it feasible in the first place.

References

[1] Modisette, L., "Milking Wireless Churn for Profit," *Telecommunications, International Edition,* Vol. 33, No. 2.

[2] Arnold, S., G. A. Reed, and P. J. Roche, "Wireless, Not Profitless," *McKinsey Quarterly*, 1999, No. 4, pp. 112–121.

[3] Donnolo, M., and M. Metzner, "Market Mechanics: Sharpen Your CPR Skills," *tele.com*, Vol. 4, No. 10, p. 26.

[4] Fowlkes, A. J., et al., "The Effect of Churn on Value," *An Industry Advisory*, Apr. 1999.

18
Cost Leadership

If you're not keeping score, you're just practicing.

—Vince Lombardi

Some telecommunications providers will analyze the market and conclude that the segment they want to serve is simply a commodity market. At present, most of the telecommunications market exhibits commodity characteristics:

- Customers perceive little differentiation between suppliers.
- Customers churn primarily to obtain a lower price.
- Customers can substitute services easily with those from competitors.
- Customers have ready access to price information.

While it is likely that telecommunications service providers will enthusiastically create brand differentiation as the market matures, the majority of competitive services are still near-commodities. Providers of undifferentiated services succeed through cost leadership. Cost leaders need to make sacrifices and often transform their corporate cultures, but this path to differentiation is practical and sustainable. Cost leadership is a financial strategy, inasmuch as the service provider's lower costs, and therefore profitability, are relatively invisible to the customer.

An alternative strategic approach is the best-cost producer strategy. This strategy combines an emphasis on low cost and a focus on differentiation. The result is that the best-cost provider probably does not have the lowest cost among its competitors. If it can identify and serve customers to whom the points of differentiation are visible and more important than the price difference, it can succeed. Ideally, the best-cost producer can compete with lower-priced rivals by offering customers additional value and compete against high-priced providers through lower prices. This strategy requires an additional focus in another path to differentiation, such as customer service, pricing innovations, or a unique business model. Bell Atlantic has actively pursued the best-cost strategy.

18.1 Characteristics of cost leadership

To maintain the status of low-cost provider, a telecommunications service provider needs to demonstrate most, if not all, of the following characteristics:

- *Preferential access to supplies:* The low-cost provider examines every element of the cost profile for opportunities to lower costs. While offshore manufacturing is not ordinarily available to domestic service providers (other than using suppliers with offshore facilities), offshore labor services are. Overseas data entry can be a cost-saving opportunity, especially to a facilities-based service provider with international network facilities that significantly reduce transmission costs. Service providers with commitments to domestic labor contracts will have a cost disadvantage against competitive service providers that are not bound to similar obligations. Furthermore, the nature of telecommunications, fostered in fact by the competitive market, will create more opportunities for competitors to redirect traditional labor tasks to inexpensive foreign or domestic services or simply to automation. On-line provisioning and flow-through operations are among the first examples of automation, which will eventually, if not already, reduce costs significantly for some service providers. VoIP telephony and telecommuting have already relaxed the boundaries between corporate facilities and other places. This has already resulted in an increase in hourly and freelance labor and an expanded market for service providers who are skilled at managing this new work model. Because these new sources of labor are technology-based, they will undoubtedly increase in capability and ubiquity.

Scale will prove to be a requirement for the low-cost telecommunications service provider, whether facilities-based or a reseller. Facilities ownership in general is preferred to optimize the cost profile, to capture the advantages of economies of scale and maintain control over asset utilization. For resellers of some or the entire service portfolio, low-cost opportunities accrue with volume. The long-distance reseller that negotiates billions rather than millions of minutes with a single supplier will command a more favorable rate per minute.

- *Low overhead:* The cost leader keeps administrative costs at a minimum. The low-cost provider sends a message to the entire work force and their business contacts by maintaining austere offices and other accouterments. Bell Atlantic disbanded its executive dining room and some other senior management privileges to save their costs and to communicate the seriousness of its best-cost program throughout the organization. Low-cost providers also demonstrate the courage to eliminate employees that do not provide better-than-average return for their salaries and other costs. Maintaining a low overhead is a constant activity, and it is difficult for both telecommunications giants and for small players. The challenge to the largest companies is that cost management is a detailed, pervasive task, and while senior management tends to focus on aggregated costs, true cost control occurs when all supervisors and first-line managers work together to control costs. For the service providers that have succeeded in cost reduction programs, a most important element of the program is often culture change. Small providers have to reduce their cost profiles without the advantages of economies of scale. Furthermore, their growth creates some new costs in a stair-step fashion. This concern is higher for small providers: first, because their growth is typically higher than that of incumbent providers, and second, because any addition—a new division, a new office, or a new information system—represents a higher proportion of overall cost. An added layer of management or a new headquarters facility will create overcapacity and associated extra cost for a significant interval. Until the service provider grows sufficiently to fill out the new administrative overhead, reducing other costs will need vigilant attention to compensate.

- *Employee low-cost focus:* Employees of the cost leader demonstrate characteristics that sustain the low-cost profile. They maintain a sense of urgency; they recognize that the cost of any activity rises as it increases in duration. Employees are most cost-effective when their responsibilities require repetitive behavior, a focus on quantity, minimal responsibility, low flexibility, a narrow skill set, and little ambiguity. This represents additional evidence that the service provider committed to cost leadership needs to assess every element of the organization and make sure that other strategic objectives do not conflict with the overall cost leadership goal. For example, employee empowerment is effective for certain paths to differentiation, but it is a needlessly costly management technique for a service provider aspiring to be the cost leader. Indeed, cost leadership is largely incompatible with many of the other paths to differentiation when those paths increase the cost profile in the expectation of commanding high prices for differentiated services. On the other hand, service providers can maintain a low-cost profile and gain other elements of differentiation when they exploit technology, scale, and a fixed cost structure to ensure low costs without sacrificing capabilities.
- *Minimal margin:* Largely, the low-cost provider anticipates high volumes and low margins. For all service providers, and particularly for those in commodity markets, above-market margins (all other costs among competitors being equal) invite competitors to lower their prices and gain market share. While some cost leaders do maintain slightly higher-than-average margins compared to their competitors, any slight disparity might be a sign of differentiation rather than superior costs. This is especially true if one service provider cannot maintain superior access to low costs without competitors imitating their techniques and closing the gap. AT&T had a temporary cost advantage when it launched its Digital One Rate wireless service, because no other major provider had a national wireless infrastructure and a nationwide wireline long-distance network. Rather than

try to maximize its margins, AT&T revolutionized the price structure and made the service more affordable than any of its rivals' offerings. The result was that AT&T amassed significant market share before its competitors were able to retaliate with their own versions of the service. By the time competitors launched their competitive services, AT&T had established a brand, amassed a customer base, and created some distance from its competitors as it improved its own business processes in managing the service.

- *Long production runs:* Low-cost service providers maintain long production runs and high asset ownership. In the telecommunications industry, this creates a dilemma for the service provider that needs to minimize cost without maintaining obsolete equipment. Service providers intent on cost leadership set long service life as a priority for their technology purchases. Besides the obvious lower costs associated with depreciation and the avoidance of new purchases, long equipment life cycles (and information systems and business processes) create a learning curve effect. Employees do not require retraining for new installations, so their productivity remains high. One decision criterion for the low-cost provider is to ensure that systems are scalable (so that the business can grow without changing vendors or equipment lines) and that the suppliers demonstrate a commitment to improving their services and equipment in a smooth and inexpensive migration. For the reseller, the cost structure is not as favorable as it is for the efficient facilities-based provider, but the reseller enjoys the ability to manage the obsolescence of the service portfolio by switching suppliers as new technologies (which are often less expensive) become available.
- *Target the average customer:* The cost leader is willing to forego the high-volume, high-maintenance customer in favor of serving the mass market. This service provider's portfolio consists of basic services that derive from low costs. Customization is only available when it is inexpensive to provide. The mass-market customer does not want to spend a great deal on customization anyway. The competitive danger is that competitors can readily invest in technologies that create minimal customization and render the low-cost provider's portfolio less attractive. When R&D efforts of competitors result in a disruptive technology, the low-cost provider is at more risk. The service provider striving for cost leadership needs to seek

out low-cost distribution networks that boast superior reach, to ensure that the channel reaches the most customers. Thus, the low-cost dial-around services advertise to targeted price-conscious consumers, and all of the major long-distance carriers tend to concentrate on direct marketing to entice consumers to presubscribe.

- *Process leadership over product leadership:* The low-cost provider invests in improving business processes rather than creating superior products. Cost leaders conduct reverse engineering of competitive services and products rather than invest in innovation. Service providers can also reverse engineer the business processes of competitors to the degree that they are visible. Product leadership is expensive and not especially efficient. Process leadership results in lower costs and nearly always improves performance as well. CLEC 2nd Century Communications uses just-in-time techniques to match its capital expenditures with the commencement of its revenue stream. Many facilities-based providers build out their networks and then offer services to customers. The 2nd Century architecture is highly decentralized, so that much of the network intelligence and its associated cost are on the customer's premises. Its up-front cost is 10% of that of a centralized provider (although most capital costs are delayed, not simply avoided). Note that the decentralization of network features is a process improvement rather than a product differentiation.

18.2 Maintaining cost leadership

The cost profile is an obsession for the cost leader. All decisions contain an evaluation of their impact on the cost profile and the associated competitive gain. Furthermore, cost leadership requires specific management skills and a conscientious cost measurement program. Former monopolists whose skills were generally in different disciplines will find it difficult but necessary to develop new skills to compete with cost leaders. On the other hand, incumbents have some advantages: access to capital, familiarity with asset utilization and management of large networks, and supervisory skills for the labor that is best suited for the cost leader. Skills for which the former monopolists will require time or training or discipline include process engineering skills, strict quantitative measurements and active accountability, tight cost control, and seeking and managing low-cost distribution channels.

Cost leadership requires a constant vigilance in every corner of the business. The discipline extends to pricing and integration of acquisitions. WorldCom would have worked hard to absorb the premium of the Sprint acquisition, and it would have needed to reduce other costs well below industry averages. Fortunately, the company has experience that it can apply to the integration process. Too many of the RBOCs purchased fast-growing, mid-sized interconnect companies (telephone equipment vendors), PC retailers, and software companies and immediately imposed their high-cost administrative structures upon them, eliminating any possibility that the acquisitions could maintain the formerly high profitability that made them attractive candidates in the first place. The most recent wave of acquisition activity demonstrates that acquisitions will become more complex. The earlier acquisitions of long-distance providers by long-distance providers and mergers between local service providers represented the most classical form of horizontal integration. Most of the integration activity surrounds the removal of duplication of activities and the selection of standard processes and information technologies. Increasingly, acquisitions are beginning to involve an IXC and a local service provider or providers from two different countries. Integration will become more complex, and the anticipated cost savings could be more elusive.

One method to maintain a low-cost profile is to select the target costs first and then develop a portfolio to meet the cost targets. This strategy requires the participation of virtually every business function; in fact, the business function that is not required in the planning process is a candidate for elimination, or at least close cost scrutiny. What is the purpose of a function that does not directly benefit the service portfolio? The challenge to service providers that choose to perform to a target cost level is that this approach requires considerable discipline, especially for going concerns. If the first cost profile analysis does not yield an acceptable cost level, managers then face the unsavory task of eliminating existing costs that they undoubtedly deem essential already or they would have eliminated them sooner.

Another approach to cost leadership is the management of fixed and variable costs and their relative prominence in the cost structure. While every situation is unique, management can adjust the relative proportions of fixed and variable costs to maintain a superior cost profile. In general, service providers with relatively higher fixed costs enjoy higher profits (or can lower their prices competitively) as long as volume is sufficiently high. On the other hand, there are many exceptions to this general rule. At low asset utilization, the cost advantage vanishes, and indeed a service provider with higher variable costs would enjoy an advantage over a facilities-based service provider with underutilized assets. When the service provider is uncertain what the customer volume will be, variable costs are safer than a fixed cost structure, and as volumes grow, management needs to revisit the situation. Furthermore, sustaining high fixed equipment costs in a facilities-based network risks obsolescence. Not only does the high fixed cost service provider need to maintain a high level of volume, the calculations justifying the initial capital expenditure assume that the facilities will have a known service life. If that service life is unreachable because of competitive alternatives for customers, stranding the investment or disposing of assets early is much more costly than a variable cost structure would have been. Outsourcing is one way to trade fixed costs for variable costs. Service bureaus can establish per-transaction charges for service providers that prefer their cost structure to be usage-based. Selling through agents on commission instead of a salaried sales force can create a variable profile for sales costs. Besides its advantage of creating a variable cost structure, commission plans also match the timing of expenditures for the sale to the revenues from the customer. Commission-only sales plans are not restricted to outside sales agents. A service provider can employ sales agents at a very small base salary and create a cost profile that is nearly completely variable.

18.3 Cost drivers

Conscientious cost management requires knowing which transactions create costs. Cost drivers are those activities and other business characteristics that increase costs. The concept of a cost driver arose in the field of cost accounting. Cost accounting techniques are quite common in manufacturing concerns, but until the last decade or so, service providers did not use them to improve their own productivity. The relatively recent interest in reengineering and business process redesign has sparked a corresponding attraction to cost accounting methods in service industries. At the same time, manufacturing enterprises have moved toward service-based measurements as their businesses have become more automated and their blue-collar labor has diminished in favor of white-collar knowledge workers.

Some cost drivers are structural: The facilities-based provider has maintenance costs driven by ownership of a network. Other drivers are transactional; a sale by an agent creates a commission expense. Some are apparent; enterprises with high variable costs find business volume to be a significant driver. Others are surprised to discover that cost drivers other than volume make up a large proportion of costs. Certainly, no service provider would set a goal to reduce volume in an effort to reduce its costs, but reducing many other cost drivers can help to meet the objective of cost leadership. Churn is one of the best examples of a cost driver. The difference between the loyal customer and the replacement customer (assuming that the service provider's growth replaces churning customers) is the cost of customer acquisition and termination. For even the most ordinary sales transactions, the cost of customer acquisition can be at least hundreds of dollars. The cost of retaining a customer is less per year than the turnover associated with churn. The most rudimentary analysis can reveal the size of investment that a service provider can make to reduce churn and reduce the overall cost profile.

For DSL provisioning, one key cost driver is the number of truck rolls to the customer's site, dispatching technicians to perform manual tasks. A typical installation requires two to three truck rolls at hundreds of dollars each. Unnecessary truck rolls occur when the service order is wrong, when the service provider's research is inadequate to screen out customers whose facilities do not qualify for DSL, or when insufficient maintenance causes unnecessary service failures. The average DSL installation requires 1.5 unnecessary truck rolls [1]. New DSL standards (Universal Asymmetric Digital Subscriber Line or G.lite) will lower costs partly through the elimination of most or all truck rolls to the customer, among other cost improvements. Indeed, the major benefit of the new standards is that they reduce installation complexity and cost.

Another important cost driver is network service interruption. Service outages create repairs, customer dissatisfaction and potential churn, possible penalties due to service-level agreements or other contracts, and unexpected corporate communications, at best. Service providers that choose not to compete based on service quality need to evaluate the proper balance between network reliability and the costs of recovery.

The pricing of services will prove to be a controllable cost driver as competition continues to intensify. The movement in the United States and Europe to reduce the amount of service measurement will create simpler bills, lower printing costs, and lower customer service costs. The service provider that immediately writes off the disputed charge for a single call could potentially save more than the network expense in customer care time, not counting the cost of customer dissatisfaction. According to Boston Consulting Group, about 75% of customer service calls relate to billing, many of which are disputes [2]. This represents a large cost reduction opportunity, as the customer service function represents 7–10% of a usage-based service provider's total costs.

To identify and control cost drivers, service providers can map each of their business processes. Through a variety of analyses, management can identify the most significant cost drivers and the most controllable cost drivers. The solution might involve redesigning business processes, changing distribution channels, or providing customers with clearer bills.

18.4 Profit centers and cost centers

Changing cost centers into profit centers is worthwhile whenever it is feasible. The conventional view of the difference in the two structures is that managing cost centers focuses on cost reduction, while profit centers can create income. Measuring the two entities differs substantially. Enterprises expect profit centers to produce income (revenues exceeding its costs). The goal of a cost center is to manage to a budget and spend (have a loss) as little as possible.

One obvious candidate for transformation is the call center. Whether the call center acts as a customer service interface or a help desk, service providers can refocus the call center, if it fits the overall strategy. New technologies enable the call center to provide value-added services to customers or take advantage of cross-selling or up-selling opportunities. Suppose a customer calls to question line items on a bill. As a cost center, the service representative can remove the disputed items or explain their presence to the customer. Either of these activities creates cost and possibly customer satisfaction that offsets the cost. As a profit center, the service representative can take a more active approach, reviewing the customer's specific service needs, changing the service bundle, and potentially up-selling new services that meet the customer's requirements. This approach obviously creates more cost, but it also holds the potential to increase revenues at a level surpassing their costs.

Similarly, the help desk call center can respond to technical support questions reactively, or the technicians can help customers learn about training seminars or service add-ons that facilitate the customer's operations. Another example of a cost center with profit potential is the service provider's Web site. As the service provider establishes on-line commerce, revenues from the site (and potentially outside advertising or affiliated sales) can overtake the costs of maintaining the presence on the Internet. On the other hand, customers calling for support and not sales could resent the representative guiding the conversation towards sales. Turning all support functions to sales can undermine customer satisfaction, creating high cost and indirectly causing churn.

For those functions remaining as cost centers, the telecommunications service provider needs to perform another assessment: whether to outsource the function. If a function clearly cannot generate profits, then management of the function must hold costs to the lowest possible level. Either the service provider should apply resources and focus to ensure that costs are low or find a low-cost provider and outsource the task. Figure 18.1 demonstrates the transition from a cost center to a profit center.

18.5 Cost measurement

Telecommunications service providers are adept at measuring network performance, but measuring other aspects of operations has not generally commanded the same priority. As information technologies and knowledge workers have captured a larger share of the overall cost of the organization, performance measurement is occupying a more prominent position. Especially for the cost leader, measurement is crucial. The service provider needs to develop metrics and objectives that have several characteristics:

- *The metrics and objectives need to be directly indicative of cost leadership.* It is tempting to utilize measures that are a direct byproduct of business operations rather than create new measurements and new measuring systems. Accurate cost records are the foundation of cost leadership, but costs are obscure. Accounting costs differ from operational costs. Fully allocated costs are sometimes appropriate but so are incremental costs on other occasions. Opportunity costs are sometimes relevant and sometimes they simply confuse the analysis. Metrics require management agreement in advance, but

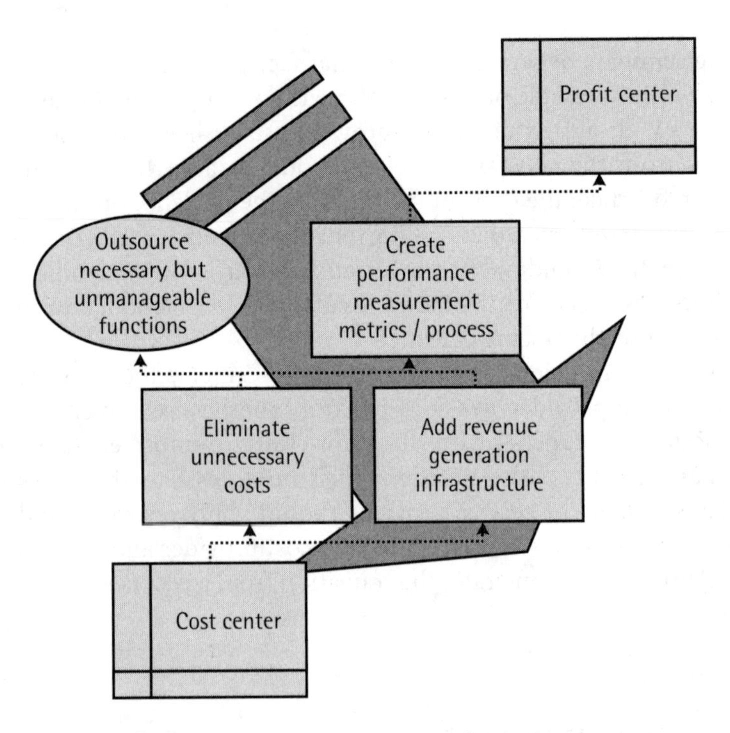

Figure 18.1 Transforming a cost center to a profit center.

they also require testing and reassessment to ensure that they measure as intended.

- *Metrics need to demonstrate constant incremental change to demonstrate continuous improvement.* Costs tend to fall in technology enterprises, and more importantly, costs tend to fall dramatically in deregulatory markets. Recent history has demonstrated to service providers that meeting goals is at best a temporary event, and goals tend to rest with the best performer in the market. Service providers that strive for continuous improvement and incremental change rather than meeting a dramatic one-time objective will create the environment needed to maintain cost leadership.
- *Management should divide metrics sufficiently to assign individual responsibility for their completion.* While many managers can agree on some metrics, assigning specific responsibility is a separate issue. Cost leaders divide the performance objectives, creating distinct

lines between organizational responsibilities, to create clear accountability for results. Cost leaders do not shrink from eliminating managers or others who do not meet objectives, assuming that competitors are at least as talented at cost reduction. Managers accepting objectives accept the specific result, the accompanying consequences, and the required time frame.

- *In a competitive market, measuring costs is the foundation of developing profitable prices.* Historically, telecommunications pricing did not reflect the cost structure of the provider. Besides the well-known subsidies flowing from long-distance to local service, from businesses to consumers, and from urban areas to rural areas, other cost recovery mechanisms obscured cost structures to customers. Service providers amortized fixed costs such as installation labor to reduce the first costs for customers. Telecommunications service providers imposed fees for new technologies such as tone dialing and never eliminated them when the revenues were no longer necessary. While some accounting or pricing nuances will still be essential in an intensely competitive market, overlooked residual pricing strategies can lead to competitive arbitrage, as profitable customers flock to services without subsidies and unprofitable customers have nowhere else to go.

18.6 Supporting tools

Cost leadership requires constant attention, and vigilance requires tools. New tools have taken much of the drudgery and human error out of cost accounting and analysis.

Activity-based costing

The most widely used costing analysis technique is activity-based costing (ABC). ABC developed as a response to traditional cost accounting methods. Traditional accounting practices tend to smooth out fluctuations in cost, which is valuable to investors for comparing companies but can be misleading to operations managers and senior executives for managing their own operations. Accounting techniques also focus on a single cost component rather than the complex costs now involved in manufacturing and service industries. ABC is extremely effective for companies with diverse product lines, many customer segments, intense competition, or falling prices, so it fits the telecommunications industry well.

An ABC analysis defines the business into categories of cost and business processes. The analyst defines a list of activities within the business processes. Employees complete questionnaires (or supervisors complete questionnaires for workers) that allocate their work time among the activities listed. From there, with the help of software support, the technique calculates the cost and dispersion characteristics of each business process or other defined operation. ABC can assign costs to specific services or to customers. It is not unusual to learn that a large number of services or customers are not profitable as expected, after considering all of its costs.

ABC focuses on the specific activities associated with operating the business. The technique allows for significant analytical flexibility, so it can answer questions such as "how much does the average billing complaint cost?" or "have we assigned employees with the proper skills to manage the provisioning process?" ABC is useful in identifying high- and low-value activities, managers performing activities beneath their skill level (and at excessive cost), activities fragmented among too many employees, and other incongruities. A thorough ABC analysis can result in significant cost savings in areas that are otherwise invisible to management: cross-functional discontinuities, excessive costs of management, and disproportionate costs for problem-solving that can be eliminated through additional training or other front-end activities.

ABC analysis is detailed and time-consuming, and its value is often proportionate to its comprehensiveness, within reason. Still, the largest expense and commitment are in analyzing the data. Fortunately, many PC-based software packages are now available that relieve much of the burden. A service provider committed to maintaining cost leadership will conduct activity-based costing studies regularly and pay attention to their results.

18.7 Role of information technology

Information technology plays a prominent role for cost leaders. Cost leaders innovate rather than automate with information technology. Cost leaders restructure their entire businesses to maximize technologies with proven cost savings. Information technologies, coupled with management techniques such as just-in-time processing or EDI, can revolutionize business processes while realizing dramatic cost reductions. The examples of innovation through technology and design generally come from new entrants to the industry. Skeptics of incumbent service providers would undoubtedly attribute their innovations to the more entrepreneurial culture of newcomers and their hunger for customers, but that is only partly true. Another reason that new entrants can innovate readily is that they are unencumbered by the network architecture, business processes, mass-market customer base, and remnants of regulation that limit the incumbent providers. Incumbents thrust into a competitive arena have already proven that they can be innovative with technology when similarly unconstrained. Technology also offers incumbents an opportunity to diminish the role of labor in their organizations, historically a source of high cost.

Information technology has also facilitated the business-to-business e-commerce infrastructure, which reduces purchasing costs by eliminating intermediaries, offering bulk purchasing power to small service providers, and automating manual activities. Furthermore, technology fosters the record-keeping and analysis required on an ongoing basis to ensure the accurate and prompt measurement of costs, so that management knows soon enough about areas of concern to do something about them. Besides the capture of operational and financial information on an ongoing basis, performance measurement software such as support for ABC and project management enables planners and managers to analyze historical data and simulate a potential improved cost profile.

References

[1] Aaron, J., "How To Put the Brakes on DSL's Costs," *America's Network*, Vol. 104, No. 1, p. 37.

[2] Hixon, T., and C. Moffett, "The End of Time," *tele.com*, Vol. 4, No. 18.

19

Funding Strategies

Telephone companies should not raise their rates because their networks are already paid for.

—Consumer advocate petition

While innovative funding strategies do not normally differentiate a service provider to its customers, they can make a difference in service deployment, capital costs (and thus pricing), and other fundamental characteristics that are quite visible to customers, not to mention shareholders. Therefore, the telecommunications service provider that innovates through financing its operations and managing its financial portfolio will hold a competitive advantage over an equally managed service provider that fails to do the same.

19.1 Financing alternatives

Industry observers pay close attention to the startup investments of industry leaders and, on the other end of the spectrum, to the shoestring operations that become legends with a brand new business model that creates millions or billions of dollars in shareholder value. In reality, most of the startup activity in the telecommunications services industry falls into the middle of the continuum. The primary participants are CLECs, niche providers, and regional network providers. All of these aspirants, and especially those intending to offer facilities-based services, need capital. Going concerns also need funding routinely for new projects, growth, facilities upgrades, acquisitions, or modernization.

There are two basic ways to finance a new business or to fund business growth: debt and equity. Debt is simply a loan from a bank or another lender. The enterprise borrows a certain amount at a certain rate and repayment schedule. Debt can represent more than half of the capital structure of a telecommunications service provider. Incumbent telecommunications providers often held a high ratio of debt to equity. In a monopoly environment, debt is inexpensive compared to equity funding. Monopolists are almost never in danger of being unable to pay back their debt requirements. Overall, debt financing helped to keep rates low. In a competitive environment, newly deregulated providers are revisiting their debt ratios and reducing the amount of debt they hold. As competition intensifies, incumbents will have more business risk, and their cost of debt will rise accordingly. New entrants to the telecommunications service business will undoubtedly pay more for debt financing than the incumbent providers do in the same deregulated marketplace.

Equity financing enables investors to benefit from the profitability of the business. Investors also risk losing their investments if the business fails. Equity is more expensive than debt, primarily because investors are willing to take on more risk than lenders are and they expect reimbursement for their risk. Lenders are also first in line for repayment if an enterprise fails, so investors also take most of the burden of failure. Too much equity funding can dilute shareholder value, but too much debt creates a payment commitment that might be too high for the business to sustain during slow business cycles. Another advantage of equity funding for telecommunications service providers and especially Internet companies is that equity in promising businesses is in demand, before and after they go public. According to Deutsche Banc Alex Brown, the value of 25 Internet-related companies that went public in 1999 had gone up an average of nearly 400% by year's end [1]. To capture equity funding, new entrants presently hold a slight advantage over incumbents, as risk-tolerant investors who expect high returns have made a great deal of capital available to technology startups.

Venture capital is a common method for acquiring equity financing. Figure 19.1 demonstrates the considerable interest in the investment community for telecommunications opportunities. A 1999 Pricewaterhouse-Coopers/*Network World* study predicted a more aggressive estimate of total venture capital investment in network companies at $16 billion, triple that of the previous year.

Not every telecommunications entrepreneur will prefer venture capital financing, and very few will win it. On average, venture capital firms receive more than 1,500 business plans per year and invest in only about a dozen [2]. Generally, venture capitalists seek out enterprises with the intention of taking them public or selling them outright in a specified period of a few years. Entrepreneurs that desire to create a business to bequeath to their heirs should look to other forms of financing.

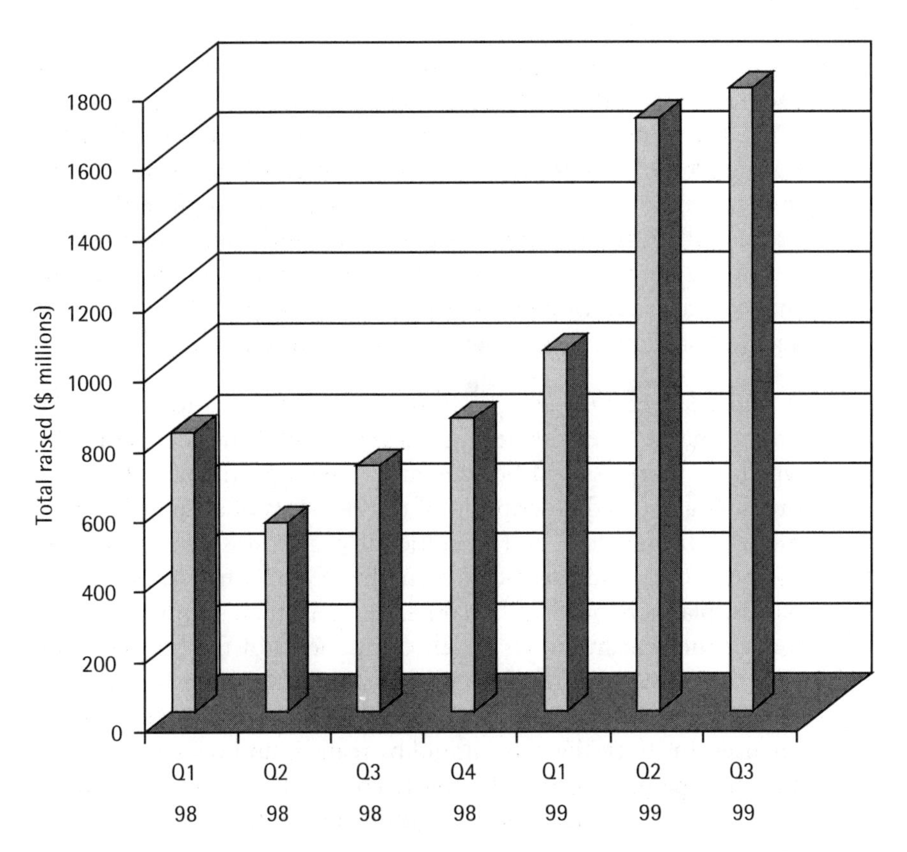

Figure 19.1 Growth in venture capital.

Seed financing assists an inventor or entrepreneur to develop and prove a concept or prototype. Startup financing enables the new company to complete product development and initiate marketing. First-stage financing funds full-scale manufacturing, if required, recruiting the management team, buying equipment, and beginning marketing and sales. Second-stage financing provides working capital to expand a going concern. In this stage, the enterprise is probably not yet showing a profit, although by this point, it probably demonstrates adequate growth and market potential. Third-stage or mezzanine financing funds the enterprise's major expansion effort. These can include plant expansion, additional marketing, and development of expanded products. Venture capitalists (and others willing to fund new businesses or business growth) look for the enterprise's business plan to satisfy several criteria, including the following:

- *A strong management team:* Investors and lenders want to see a management team with industry experience and success in previous ventures. The venture's senior management is responsible for selling the business opportunity personally to those considering making an investment, and the business founders need to recognize that this process will be time-consuming and demanding. While it is ideal for a well-rounded team to present plans to venture capitalists, a venture capital group with industry experience can fill gaps in the management skill set through their own recruiting efforts.

- *A sound business model:* The business model needs to convey the unique market need the new business will fulfill and the method by which the business will become and remain profitable. Chapter 8 provides a detailed description of the elements of a viable business model. The financial and operational projections can create best case, expected, and worst-case scenarios. Even with a variety of scenarios, planners need to be conservative in their estimates. If the management team knows which components of the business case hold the most risk, the business plan should state them clearly and identify what actions management will take to mitigate their consequences. In short, the plan should be realistic and responsible. Ideally, management needs to be able to convey the crux of the business model to a potential investor within only a few straightforward sentences.

- *Other sources of capital:* Both creditors and investors prefer that there are other sources of capital beyond whatever they plan to provide, although this is more important to creditors than venture capitalists. Creditors often ask for collateral that might or might not be relevant to the business opportunity. Venture-related collateral is problematic sometimes; telecommunications equipment and configured software have little value unless they are installed and in use in a going concern. Entrepreneurs sometimes use their personal assets, such as real estate or their own investments, as collateral against their new ventures. Knowing that there are other sources of capital is particularly important to creditors, which do not have the same opportunity for active participation in business decisions as venture capitalists and other investors often require. Most investors are willing to fund ventures without other capital sources, but they recognize the benefit of debt financing against the returns they will achieve.
- *A viable technology basis:* Many new ventures involve relatively unproved technologies. For these, the enterprise management needs to demonstrate the merits of the new technologies, confirm that the technology will work in the current telecommunications environment, and handle concerns about its development risk if it is not already in the marketplace. For new technologies that also offer new services (or existing technologies retrofitted to provide new services), the team might also need to define the market segments for the new services and validate the anticipated sales price and demand. The team can also describe the eventual portfolio of products that the new technology makes possible. If the technology involves significant switching costs, such as replacing software or reformatting content, the team should develop a feasible and inexpensive migration path for customers. This might require additional technology development.

Another significant difference between venture capital firms and creditors is that venture capital firms expect a significant return on their investment, from 50% or more in early stages of profitability to up to 30% in the later stages. Creditors only care about interest payments and have little concern whether an enterprise is enormously successful, as long as it remains profitable for the term of the loan.

Many of the telecommunications leaders maintain in-house startup operations that they fund themselves. Legendary success stories of huge businesses incubated internally, such as IBM's development of the PC, have prompted telecommunications service providers to build and support their own in-house operations. One reason that IBM's incubator worked is in its physical and intellectual separation from the conservative approach of its industry-leading parent company. The decision to create an open architecture for which third parties could develop software applications was innovative and almost heretical. The decision to open the architecture by the manufacturer most expected to succeed in the PC industry enabled the explosive growth of the industry (and plug-compatible competitors). The semimonopolistic IBM management would have found it difficult to plan to take a secondary or shared leadership role or to cannibalize its still-leading mainframe business. The IBM senior management team might fear that a standard software interface might reduce its opportunities to differentiate its own software products. Nevertheless, leadership manufacturing and marketing in the conventional IBM style might have garnered the company a similar share, but more likely a separate proprietary standard might have excluded the company completely in the long run and hindered software market development. IBM now enjoys a significant and profitable market presence in PCs and desktop software, and it used this experience to create a significant presence in the emerging e-commerce market.

There are no announced triumphs for in-house telecommunications startups yet (wireless spin-offs would not qualify), but competition is still immature in most areas. There are some challenges facing the large corporation that are not as apparent to their entrepreneurial counterparts. The large enterprise runs little risk of ignoring its incubator and more risk of overfunding it. An independent startup operation with scarce funding forces the founder to impose self-discipline on the level of expenditure and the business model itself. Corporations that fund new projects from a limitless account tend to overdesign the product, include too many features, and extend the range of products before going to market. An early market launch can identify design flaws, user-friendliness requirements, and new features soon enough in the design process to be easy and inexpensive to correct. Consultancy McKinsey recommends a milestone-based approach for internal ventures that simulates a venture capital model [3].

19.2 Competing in the investment community

Because the investment community is becoming as competitive as customer markets, telecommunications service providers have taken innovative approaches to differentiating their enterprises to sources of funding, often by the same branding techniques employed for customers. Public companies are targeting investors by advertising in consumer media. Nortel and Lucent are becoming as recognizable as MCI WorldCom and the local service providers in consumer-directed television advertising. As these ads are not cost-effective if they are targeting telecommunications equipment buyers, it is fair to assume that their intended audience is investors. Even MCI WorldCom's campaign to sell its On-Net service is probably creating interest in the company as a whole more than it sells enterprise networks to undifferentiated television viewers. Competitors that are less well known than the industry leaders often generate the interest of the investor markets on days that their CEO appears on 24-hour cable financial networks.

Some publicly traded telecommunications providers have cut their dividends to investors, which provides funding without going to the marketplace. Moreover, the booming U.S. economy has cut the dividends as a percentage of the share price, whether or not the service provider made an active decision to do so. The higher share prices mean that service providers will be able to issue equity more easily than in the past, as long as investors are willing to continue to bid the price higher. In any case, the trend toward lower dividends or none at all is likely to continue, as the newer market participants—cable providers, competitive long-distance providers, and new entrants—generally pay no dividend at all.

The shareholder base will undoubtedly change to reflect the new investor's need for value growth instead of consistent dividends. Risk-averse investors, once the mainstay of the shareholder group, will probably migrate away from the telecommunications service providers into safer investments. The increased risk in competitive markets, with assistance from a generally prosperous economy, will move these investors to alternative investments with similar rates of return to traditional utility investments. Furthermore, the sheer number of well-known local service providers will dilute this group as potential investors. When investors choose to invest in their financially stable local telephone company, they will soon have a choice of several equally attractive neighborhood investment alternatives, when SBC, Verizon, and Qwest are all present, for example, in what was once BellSouth's territory.

Suppliers have been willing to finance equipment sales for emerging service providers. The service provider benefits from the below-market interest rate, and the supplier generates an income stream and the promise of additional purchases as its customers' businesses grow. The customer also benefits from the interest and technical support that the lender/supplier will provide to ensure that the equipment operates effectively and generates the revenues and profits to pay back the investment.

Many new entrants with innovative business models benefit from the investments of the leading service providers, which use the investment as a diversification strategy and occasionally a source of acquisition candidates. Data CLECs, for example, are a favorite investment focus of telecommunications service providers. The other benefit to the telecommunications giants is that their investments serve as a form of targeted research and development expenditure. If they recognize that small companies can provide a startup function more effectively than the entrenched service provider could accomplish on its own, both parties benefit. The new entrant has superior access to equity capital, and the incumbent provider has the opportunity to earn exceptional rates of return or gain access to new technologies when they become available.

19.3 Releasing shareholder value

The goal of any publicly owned telecommunications service provider is to create shareholder value, and a similar goal is applicable to privately held service providers as well. Shareholder value is indefinable. Stocks rise when investors believe that they will rise even more. The service provider has limited control over the expectations of investors, because other factors—the economy in general, the performance of competitors, and specific industry characteristics—also influence investor perceptions of the value of its equity. Still, the telecommunications service provider that wishes to maintain its standing with the investment community must focus on value creation.

Value creation is stronger in the telecommunications industry than other segments in the robust U.S. economy. According to consultancy McKinsey, the telecommunications industry created $1.5 trillion in market capitalization (shareholder value) between 1988 and 1998 [4]. Its research also determined that more than half of the new value arose from companies not originally part of the original Bell System.

The first step, and one that a service provider might overlook, is to maintain a monitor on financial performance from the perspective of the investment community. Mercer Management Consulting suggests that telecommunications service providers calculate the ratio of a company's market value to revenue (MV/R) [5]. The MV/R ratio provides a valuable benchmark among various competitors. The service provider can calculate the ratio routinely to ensure that the direction of change is favorable. According to Mercer, a company with a ratio higher than 3 is in a state of "value inflow" and "value outflow" when less than 1. Value stability occurs when the ratio falls between 1 and 3. Mercer also notes that incumbent providers tend to be in a state of value stability, while emerging players are in value inflow.

Increasingly, telecommunications service providers are using financial instruments as well as operational ones to increase overall shareholder value. According to Broadview International, the number of global telecommunications mergers and acquisitions rose from 603 to 834 between 1998 and 1999, and their value increased from about $250 billion to more than $680 billion (about 175% higher, mostly due to European transactions). At least 33 U.S. mergers since the Telecommunications Act totaled more than $492 billion [6]. Much of the industry consolidation has involved the largest service providers. It is plausible that the second wave of consolidation will occur when the leading service providers fill out their service portfolios and geographic coverage by purchasing CLECs and other new industry entrants. SBC is under a regulatory requirement to build local facilities in 30 markets on an aggressive schedule. Acquiring a local CLEC in a desired market would meet the requirement and reduce the management required if SBC chose to build out on its own. Furthermore, acquisitions, while more costly than buildout, most often provide an instant revenue stream and potentially profits as well. Other RBOCs, similarly constrained, will undoubtedly begin to compete for candidate CLECs to acquire. Indeed, CLECs demonstrate through their own investments and operations whether their intention is to build a going concern or build an acquisition candidate. An alternative is to create a partnership whereby a national service provider and a local facilities-based CLEC interconnect and jointly market their services. This arrangement can add value to the portfolios of both partners and can be the precursor to a full merger.

Not all local service providers are competitive new entrants to the industry. Many of the more than 1,000 incumbent service providers that are not industry leaders (once known as independent telephone companies because they were not affiliated with AT&T) could also become acquisition candidates. Many of these incumbents serve rural areas where competition is not as fierce as it is in major metropolitan markets. A small local service provider commands a price of about $3,000–4,000 per access line [7]. One advantage to rural service providers is that they sometimes have access to favorable sources of low-interest funding. An alternative entry strategy is to purchase exchanges that round out a territory's coverage. This is especially desirable when one service provider covers an entire area except for certain gaps. The incumbent that currently serves the gap might want to recover its capital to build facilities elsewhere.

Telecommunications service providers are also creating value in the opposite way, by separating their various business units and the attractiveness each brings to the investment community. Service providers realized that divestiture or separation could release value after they watched the Bell System share prices climb considerably following their 1984 divestiture from AT&T. The birth of Lucent Technologies from its former parent AT&T also unleashed dormant shareholder value. Telecommunications service providers have discovered tracking stocks as an opportunity to release shareholder value without actually divesting the underlying business. A tracking stock is a class of shares a service provider links to the performance of a specific business unit. Usually the tracking stock represents a fast-growing segment of the overall business. Sprint PCS, the tracking stock for Sprint's wireless business, tripled in value in less than a year. AT&T created a tracking stock for the wireless business, undoubtedly with the Sprint success in mind and its own experience with the Liberty Media tracking stock resulting from its TCI acquisition. RBOCs, concerned that their images do not invite risk-oriented investors, have considered creating tracking stocks for the less traditional portions of their portfolios such as wireless, international businesses, and Internet-related enterprises. The tracked company attracts new investors that would not consider an investment whose growth and profits are nearly invisible within the context of the entire enterprise, and the tracking division gains from the ability to draw on the parent's credit. There are assorted challenges to managing an enterprise with a tracked unit, and the concept is relatively new, but the result for telecommunications service providers so far has been worthwhile.

Telecommunications service providers might also use restructuring to realize accumulated value. If the service provider owns a technology or a brand, it might consider licensing it through a joint marketing or licensing arrangement. If the technology is not integral to the service provider's present value proposition, or if it holds more value to a potential acquirer than within its current quarters, an outright sale might increase the seller's portfolio.

19.4 Return on investment strategies

While most acquisitions or buildouts occur because of deliberate strategies or the materialization of specific opportunities, initiatives resulting from a portfolio risk assessment can improve overall return on investment (ROI). For a facilities-based local service provider, conventional wisdom implies that the market to target is the business market. Unfortunately, the margins in the most desirable markets are often thin. The service provider seeking value can round out the portfolio with buildouts in markets with different competitive characteristics and higher margins, such as urban residential markets (which generally have lower cost than other consumer markets) or rural markets (featuring less intensity of competition and sometimes a guaranteed rate of return).

Some regulatory rulings concerning local competition will improve the return on investment for service providers reselling the services of incumbents. One analysis comparing unbundled resale determined that the cost of reselling at a discounted rate was more than 40% higher than the same elements in an unbundled platform [8]. Similarly, the FCC's line-sharing decision substantially reduced resale costs for data CLECs providing DSL services. Regulators are intent on removing competitive barriers, so they will undoubtedly continue to issue rulings that eliminate regulatory remnants that unnecessarily reduce ROI for new entrants.

Service providers can also manage their financial performance through traditional financial techniques. One very significant cost element is depreciation. Companies divide the total cost of an asset that will last for more than a year to match its costs (by recognizing accounting expenses) with its use. Depreciation is a significant concern to facilities-based service providers, whose businesses are typically very asset-intensive. As technological advances continue to drive the industry, the service life of switching and transport equipment has decreased considerably. The equipment still operates for decades; it simply does not meet market needs for much more than a few years. Regulators historically used long service lives as a technique to minimize expenses and the resulting service prices, and telecommunications monopolies controlled the operational life of facilities. Competition has changed the amount of control any single provider will have on the introduction and decommissioning of technology. Incumbent service providers are discovering that their historical depreciation schedules can prevent them from recovering their network investments when the facilities become obsolete. Technology Futures conducted a study that provided evidence that by using historical depreciation rates rather than competitive life spans for telecommunications equipment, regulators are overstating equipment lives by an average of 173% [9]. Service providers with installed facilities can improve their long-term financial performance by aligning the book values of their networks with a more realistic assessment of their usefulness. Furthermore, companies considering the acquisition of facilities-based providers need to apply due diligence to both the accounting view of the candidate's infrastructure and an objective inspection of the physical plant.

Revenue assurance is another area that service providers can improve financial performance. Deloitte & Touche released a study that concluded that revenue leakage could reduce revenues by more than 11% [10]. Revenue loss can occur within most customer-facing business processes, including product development, sales and marketing, customer care, provisioning, and billing. Delays or errors in the ordering process can postpone the receipt of revenues or drop them altogether from the billing process. Manual processing anywhere in the business cycle can introduce errors and lost revenues. Overcharged customers are much more likely to report errors than their undercharged counterparts, so service providers need to take the responsibility for revenue assurance. Similarly, telecommunications service providers can improve their collection rates by outsourcing revenue assurance functions that are likely to remain at a low priority, such as collections management. Service providers that use outside services to reduce bad debt often pay only a percentage of recovered debt to the service bureau. Other service providers will sell their accounts receivable outright at a discount, eliminating bad debt and gaining the immediate use of the collected funds.

Whether targeted or unanticipated, many incumbent service providers will need to adjust the sources of revenues in their portfolios. Andersen Consulting has suggested that local service providers could lose 15–30% of their customers each year after deregulation is complete. AT&T has stated that it expects its combined consumer and business long-distance businesses to decline from 73% of total revenue in 1998 to 30% in 2004. Besides the obvious entry of local service providers into long distance and vice versa, diversification into other geographical territories or other businesses presents an opportunity to increase overall returns and reduce financial risks. American and European service providers are eyeing each other's territories and have created cross-border partnerships such as those between AT&T and British Telecom and between KPN and Qwest. A Deutsche Telekom overture to acquire Qwest and US West failed, but more acquisitions are inevitable. European and U.S. companies have made significant acquisitions and investments in South America and Asia, and there is interest in serving markets in Africa and other underdeveloped areas that currently support little or no telecommunications infrastructure. The investments in these areas, which once focused exclusively on incumbent operators, now include investments in new access providers, ISPs, and portals as these markets are undergoing their own transformations.

19.5 Supporting tools

Tools supporting funding strategies are among the most sophisticated models available. They support telecommunications service providers, and they are vital to the financial institutions that advise service providers and investors alike. While many tools can operate through desktop software, outside advisors will be most helpful in applying the proper analytical approach and interpreting the data.

Portfolio analysis

Portfolio analysis refers to a set of tools that calculate the value of the elements in a service provider's portfolio. Portfolio analysis is often a part of the annual business planning process, but progressive senior executives review the current state of the portfolio more frequently. Portfolio analysis ensures that the funding ratios such as debt to equity and short-term to long-term are in an acceptable balance. It also reviews the performance of each line of business to ensure that each division is contributing its share. Like any investor's portfolio, service providers will find that some divisions grow more quickly or are more profitable than other divisions. Without adjusting the mix of businesses routinely, a service provider could risk creating a downturn in overall profitability or might ignore a worthwhile business opportunity that would rebalance the portfolio. One approach to portfolio analysis reviews each line of business in terms of its market position and the state of its industry. The analysis might compare the service lines in terms of their proportionate revenue. A service provider could realize that a line of business has become an extremely significant portion of the revenue stream. If the service line were emerging, management would probably be willing to continue to invest in the service line. It is also possible that the proportion of revenues is growing because other service providers are exiting an unprofitable business and that the absence of profits is damaging the overall company returns. If the industry is strategically significant or if management anticipates high returns soon, the enterprise will take different actions than it would if the industry were in decline. Moreover, management needs to ensure that the portfolio allocation—to businesses with stable profits, to businesses that will provide growth, and to investments with high risk—maintains its proportionality as ongoing operations modify the various components of the analysis.

Shareholder value analysis

Shareholder value analysis, like portfolio analysis, can use several tools. Its goal is to arrive at a measurement of shareholder value, that is, what the business is worth to its owners. Some of the tools that corporations use for other financial analyses are inadequate in that they focus too much on short-term profitability or that they rely excessively on historical data. Measures that calculate dividends or profits but ignore stock appreciation produce an incomplete view of shareholder value. Traditional methods that use profits as a key measure are not satisfactorily comparable between companies; companies may interpret accounting rules uniquely or file their financial reports in countries with varied rules. Shareholder value analysis evaluates short- and long-term cash flows and other measures of value. It uses a present value approach to estimate the long-term value of a business. Management can use shareholder value analysis at the corporate level or at the business unit level. Service providers can evaluate potential acquisitions or new business opportunities alone or in the context of the overall portfolio. Shareholder value analysis is one method to link strategy and finance.

19.6 Role of information technology

Information technology plays a limited but vital role in this domain. Personal computing with standard tools such as spreadsheets and portfolio modeling assist financial planners in computing performance metrics from a variety of perspectives and developing alternative scenarios. These scenarios can range from recalculating financial statements with an alternative capital structure or developing pro forma analyses of operations after an acquisition or a partnership. On a broader level, advances in information technology have enabled funding and financial operations mechanisms such as bandwidth markets.

19.7 Summary

Telecommunications service providers will need to invest significant analysis and decision support in their strategic planning efforts. While the evolution of the competitive market remains near its beginning and volatile, the general direction of the market has become clear. Bandwidth is becoming very inexpensive on a unit basis and will probably become a commodity. Incumbent telecommunications service providers, lacking a protected territory or assigned service portfolio, will need to compete through competitive differentiation. So will new entrants, which will need to overcome the inertia characterizing a significant proportion of subscribers. Scale alone cannot ensure success, and one's own scale is only relative to one's competitors, which continue to merge and grow. Nonetheless, the telecommunications service provider that selects a path to differentiation that exploits its own competencies and provides a sustainable and profitable business model can gain a significant share of the market's growth. The keys to success include the discipline to focus on one's actual rather than desired strengths and the willingness to reduce the business scope, not necessarily scale, to a manageable and competitive level. Other success factors will undoubtedly include a willingness to change or challenge the organization, including aspects that are the most comfortable or instinctively necessary. The telecommunications landscape has already changed substantially in only a few decades; in one more decade, it will change many times over. Telecommunications service providers that are willing to act, risk, and innovate will be the winners.

References

[1] Engebretson, J., “Want To Go IPO?” *America’s Network Telecom Investor Supplement*, Vol. 104, No. 2, pp. 8–14.

[2] Mason, S., “Winning Venture Capital Financing (It’s All in the Planning)," *Telecommunications*, Vol. 33, No. 3.

[3] Clayton, J., B. Gambill, and D. Harned, “The Curse of Too Much Capital: Building New Businesses in Large Corporations,” *The McKinsey Quarterly*, 1999, No. 3, pp. 48–59.

[4] Fertig, D., C. H. Prince, and D. Walrod, “What Kind of Telco Is the Fairest of Them All?” *The McKinsey Quarterly*, 1999, No. 4, pp. 144–148.

[5] Vujaklia, P., “The New Parameters,” *Telephony*, Vol. 236, No. 6, p. 38.

[6] Hirschman, C., “Legislative Hangover,” *Telephony*, Vol. 238, No. 13, pp. 28–36.

[7] Bahr, S., “Want To Buy a Telco?” *America’s Network Telecom Investor Supplement*, Vol. 103, No. 12, pp. 8–10.

[8] Engebretson, J., “Loops Unlocked,” *Telephony*, Vol. 236, No. 6, p. 9.

[9] Britt, P., “A Depreciating Experience—Carriers Argue That Regulators Need To Rethink Accounting Procedures,” *tele.com,* Vol. 4, No. 24.

[10] Levine, S., “Do You Know Where Your Revenue Is Going?” *America’s Network*, Vol. 104, No. 1, p. 12.

List of Acronyms

3G Third-generation wireless

ABC Activity-based costing

ASP Application service provider

ATM Asynchronous transfer mode

CDPD Cellular digital packet data

CLEC Competitive local exchange carrier

CPP Calling party pays

CRM Customer relationship management

DSL Digital subscriber line

DTH Direct-to-home (satellite)

DWDM Dense wave division multiplexing

EDI Electronic data interchange

ERM Enterprise resource management

ERP Enterprise resource planning

EVA Economic value added

FCC Federal Communications Commission

GPRS General packet radio service

GPS Global positioning system

GSM Global system for mobile communications (also global service mobilization)

ICP Integrated communications provider

IMT-2000 International mobile telecommunications 2000

ION Integrated on-demand network (Sprint)

IP Internet protocol

ISDN Integrated services digital network

ISP Internet service provider

IVR Interactive voice response

IXC Interexchange carrier

LEC Local exchange carrier

LEO Low Earth orbiting (satellite)

LMDS Local multipoint distribution service

LNP Local number portability

MAN Metropolitan area network

MMDS Multipoint multichannel distribution system

MTTR Mean time to repair

MV/R Market value to revenue

NOC Network operations center

NPV Net present value

OECD Organization for Economic Cooperation and Development

OSS Operations support system

PC Personal computer

PCS Personal communications service

RBOC Regional Bell operating company

ROI Return on investment

UMTS Universal mobile telecommunications system

VoIP Voice over Internet protocol

WAP Wireless access protocol

About the Author

Karen G. Strouse is the owner of Management Solutions, a consulting firm specializing in the changing telecommunications industry. Her client list includes large and midsized telecommunications providers.

Ms. Strouse has assisted large and small telecommunications providers in developing entry strategies for new lines of business. She has conducted industry and competitive analyses to support diversification efforts and devised strategies, marketing approaches, pricing, and infrastructure to support new ventures. In addition, Ms. Strouse has helped clients to prepare business and strategic plans, develop acquisition strategies, perform competitive analyses, and improve operational effectiveness. In various industries, she has directed teams to improve business operations through the reengineering of business processes and has constructed effective organizational structures.

Her experience includes a 13-year career with Bell Atlantic, where she served in a variety of roles, including director of strategic planning. After working with Bell Atlantic, she was a consulting manager with the international firm Deloitte & Touche for five years. Ms. Strouse founded Management Solutions in 1992. Her first book, *Marketing Telecommunications*

Services: New Approaches for a Changing Environment, was published by Artech House in 1999. Her e-mail address is kstrouse@msn.com.

Ms. Strouse holds B.A. and M.A. degrees in communications from Temple University and an M.B.A. in finance from St. Joseph University in Philadelphia.

Index

Recent Titles in the Artech House Telecommunications Library

Vinton G. Cerf, Senior Series Editor

Access Networks: Technology and V5 Interfacing, Alex Gillespie

Achieving Global Information Networking, Eve L. Varma, Thierry Stephant, et al.

Advanced High-Frequency Radio Communications, Eric E. Johnson, Robert I. Desourdis, Jr., et al.

Asynchronous Transfer Mode Networks: Performance Issues, Second Edition, Raif O. Onvural

ATM Switches, Edwin R. Coover

ATM Switching Systems, Thomas M. Chen and Stephen S. Liu

Broadband Access Technology: Interfaces and Management, Alex Gillespie

Broadband Network Analysis and Design, Daniel Minoli

Broadband Networking: ATM, SDH, and SONET, Mike Sexton and Andy Reid

Broadband Telecommunications Technology, Second Edition, Byeong Lee, Minho Kang, and Jonghee Lee

The Business Case for Web-Based Training, Tammy Whalen and David Wright

Communication and Computing for Distributed Multimedia Systems, Guojun Lu

Communications Technology Guide for Business, Richard Downey, Seán Boland, and Phillip Walsh

Community Networks: Lessons From Blacksburg, Virginia, Second Edition, Andrew M. Cohill and Andrea Kavanaugh, editors

Component-Based Network System Engineering, Mark Norris, Rob Davis, and Alan Pengelly

Computer Telephony Integration, Second Edition, Rob Walters

Desktop Encyclopedia of the Internet, Nathan J. Muller

Digital Modulation Techniques, Fuqin Xiong

E-Commerce Systems Architecture and Applications, Wasim E. Rajput

Enterprise Networking: Fractional T1 to SONET, Frame Relay to BISDN, Daniel Minoli

Error-Control Block Codes for Communications Engineers, L. H. Charles Lee

FAX: Facsimile Technology and Systems, Third Edition, Kenneth R. McConnell, Dennis Bodson, and Stephen Urban

Guide to ATM Systems and Technology, Mohammad A. Rahman

A Guide to the TCP/IP Protocol Suite, Floyd Wilder

Information Superhighways Revisited: The Economics of Multimedia, Bruce Egan

Internet E-mail: Protocols, Standards, and Implementation, Lawrence Hughes

Introduction to Telecommunications Network Engineering, Tarmo Anttalainen

Introduction to Telephones and Telephone Systems, Third Edition, A. Michael Noll

IP Convergence: The Next Revolution in Telecommunications, Nathan J. Muller

The Law and Regulation of Telecommunications Carriers, Henk Brands and Evan T. Leo

Marketing Telecommunications Services: New Approaches for a Changing Environment, Karen G. Strouse

Multimedia Communications Networks: Technologies and Services, Mallikarjun Tatipamula and Bhumip Khashnabish, editors

Packet Video: Modeling and Signal Processing, Naohisa Ohta

Performance Evaluation of Communication Networks, Gary N. Higginbottom

Performance of TCP/IP Over ATM Networks, Mahbub Hassan and Mohammed Atiquzzaman

Practical Guide for Implementing Secure Intranets and Extranets, Kaustubh M. Phaltankar

Practical Multiservice LANs: ATM and RF Broadband, Ernest O. Tunmann

Protocol Management in Computer Networking, Philippe Byrnes

Pulse Code Modulation Systems Design, William N. Waggener

Service Level Management for Enterprise Networks, Lundy Lewis

SNMP-Based ATM Network Management, Heng Pan

Strategic Management in Telecommunications, James K. Shaw

Strategies for Success in the New Telecommunications Marketplace, Karen G. Strouse

Successful Business Strategies Using Telecommunications Services, Martin F. Bartholomew

Telecommunications Department Management, Robert A. Gable

Telecommunications Deregulation, James Shaw

Telephone Switching Systems, Richard A. Thompson

Understanding Modern Telecommunications and the Information Superhighway, John G. Nellist and Elliott M. Gilbert

Understanding Networking Technology: Concepts, Terms, and Trends, Second Edition, Mark Norris

Videoconferencing and Videotelephony: Technology and Standards, Second Edition, Richard Schaphorst

Visual Telephony, Edward A. Daly and Kathleen J. Hansell

Wide-Area Data Network Performance Engineering, Robert G. Cole and Ravi Ramaswamy

Winning Telco Customers Using Marketing Databases, Rob Mattison

World-Class Telecommunications Service Development, Ellen P. Ward

For further information on these and other Artech House titles, including previously considered out-of-print books now available through our In-Print-Forever® (IPF®) program, contact:

Artech House
685 Canton Street
Norwood, MA 02062
Phone: 781-769-9750
Fax: 781-769-6334
e-mail: artech@artechhouse.com

Artech House
46 Gillingham Street
London SW1V 1AH UK
Phone: +44 (0)20 7596-8750
Fax: +44 (0)20 7630-0166
e-mail: artech-uk@artechhouse.com

Find us on the World Wide Web at:
www.artechhouse.com